Inspired by Drink

Also edited by Joan & John Digby

FOOD FOR THOUGHT

INSPIRED BY DRINK

AN ANTHOLOGY

Edited by Joan & John Digby

Collages by John Digby

a toast to new friends!
To your health!
Joan Digby

WILLIAM MORROW AND COMPANY, INC.

New York

Library of Congress Cataloging-in-Publication Data
Inspired by drink:
1. Beverages—Literary collections. 2. Alcoholic
beverages—Literary collections. 3. Drinking customs
—Literary collections. I. Digby, Joan. II. Digby,
John, 1938– .
PN6071.B43I5   1988   808.8'0355   88-5228
ISBN 0-688-06922-3

Printed in the United States of America

First Edition

1 2 3 4 5 6 7 8 9 10

BOOK DESIGN BY MINA GREENSTEIN

A toast to friends

PREFACE

While poets long for the sacred fountain of the muses, the more prosaic among us long more simply for "the pause that refreshes." Every life is elevated and revived by drink. And whether we draw water from the well, crack open a coconut, drink champagne from fluted crystal, tap a beer keg, squeeze an orange, or pour another cup of coffee, our lives are punctuated by the sipping of drinks that revitalize the body and transform the human spirit.

This collection of writings is at once a fountain, vineyard, beer hall, bar, café, and teahouse. It is intended to reveal the power of drink to inspire creativity. From pure water to magic potions the poems, stories, and essays explore the diversity of beverages claimed as the creators and sometimes the destroyers of inspiration. The chapters are arranged as a beverage list, and the beverages themselves are alphabetized in a glossary at the end of the book which, in addition to containing definitions and historic information, also includes recipes for the mixed drinks mentioned in the writings so you can find out for yourself how many ounces of truth there are in authors' claims. To your health!

ACKNOWLEDGMENTS

At the *happy hour* of publication we lift our glass in thanks first to our editor, Randy Ladenheim, whose knowing hand has stirred—not shaken—these delicate spirits.

Since a cellar is only as distinguished as its vintages, we must as *sommeliers* of this collection pay tribute to the authors and translators who have allowed us to uncork their diverse and fragrant bouquets.

Finally, at the C. W. Post Campus of Long Island University, we must set up a round for the reference librarians, whose attitudes never need adjustment, even when confronted with requests for obscure information, and for the Faculty Research Committee, which has generously supported this project.

CONTENTS

♥

CHAPTER ONE
Tantalus Quenched

CHAPTER TWO
The Soul of Wine

CHAPTER THREE
Here's to Hops

CHAPTER FOUR
Blithe Spirits

CHAPTER FIVE
Drunk or Sober

Y

CHAPTER SIX
Cider to Soda

♆

CHAPTER SEVEN
Coffee, Tea, or Milk?
A Little Cocoa Perhaps!

Y

CHAPTER EIGHT
Bewitching Brews

INTRODUCTION

Every drink of water is a toast to life. We live on a planet two thirds of which is covered by water, and our own living bodies, despite their hardest bones and toughest muscles, are a staggering 70 to 85 percent water. When the pre-Socratic Greek philosopher Thales, in search of elemental matter, declared that "all things are water," he came very close to the mysterious reality of life on earth. Our whole cosmos is busy drinking, according to the Renaissance poet Abraham Cowley, who describes the cycle of nature as a revel of watery drunkenness: "The thirsty Earth soaks up the Rain"; plants nurse on the earth, the sea imbibes rivers, the sun drinks the sea and is in turn drunk up by the moon and stars. "Nothing in Nature's sober found,/ But an eternal Health goes round."

Indeed, the most common toast, "to your health," reiterates this basic connection between liquid and life. Thirst must be quenched, for without water we die. Writers have been extremely emotive about their thirsts, both literal and figurative. The aching tongue and parched throat are physical manifestations that sometimes allude to a longing for pleasure, for love, or for spiritual satisfaction. Like the mystics who write of their "dryness," the Jamaican-born American poet Claude McKay speaks of his "fevered brain" burning for "pure water . . . to wash me, cleanse me, and to quench my thirst." He seeks not merely a drink, but full immersion in the source of purity.

We can understand from his desires the utter and unrelieved agony of Tantalus who stands immersed but out of the reach of fulfillment. Tantalus dipping in the water of hell is a negative symbol of pollution. By contrast we understand every symbol of divinely pure water: Jacob's well, the water of paradise, the Helicon spring of the Greek muses, and the river "that went forth from Eden to water the garden." In our own age of polluted rivers and

oceans these images of purity are not only powerful romantic symbols of the paradise lost, but implicit indictments of human progress. How far we have come when we purchase bottled water
from natural springs advertised as "earth's first soft drink"! No
wonder so many would rather drink wine.

From the earliest grape harvests depicted on Egyptian tombs to
the libations downed by Homeric warriors who sailed the "wine-
dark" sea, tribes have praised their gods for this fruit of the vine:

> Let us praise God with this symbol of joy and thank Him for the
> blessings of the past week, for life and strength, for home and love
> and friendship, for the discipline of our trials and temptations, for
> the happiness that has come to us out of our labors. . . . Praised
> be thou, O Lord our God, King of the universe, who hast created
> the fruit of the vine.

Raising their kiddush cups, the Jews offer this thanks for their
special harvest. In Jewish prayer not water but wine and bread
stand together as the drink and food of life. When Jesus turned
water to wine at the marriage at Cana, this mysterious transfor-
mation foreshadowed the sacrament of the Eucharist in which the
wine commemorates the blood of his sacrifice.

Quite apart from mythologies, wine itself remained a mystery
for centuries. While the practice of making wine grew more sophisticated and various, the biological process of fermentation remained an undecoded secret until 1857, when Louis Pasteur
published his discovery that grape sugar is converted into alcohol
and carbon dioxide by the metabolism of live yeast cells. No vint-
ners had made better use of carbon dioxide than the producers of
champagne, who had already been capturing its sparkles in their
gleaming diamond of wines for nearly two centuries by the time
of Pasteur's discovery.

Science had simply caught up with folkways, for around the
world in tiny villages, farmers had relied on sun, soil, and local
grape stock to produce thousands of vintages distinguished by nature's birthrights. Thus the poetry of wine is a paean to nature
tinted with local color. Often it is a poetry of place names—
Amontillado, Beaujolais, Bordeaux, Bourgogne, Chablis, Cham-

pagne, Madeira, Málaga, Marsala, Valpolicella—even place names from the Napa Valley to New York State.

And if time is credited with the maturing of fine wines, there is hardly a time or a place that is not credited with an inspirational vintage. Oriental poetry represents intoxication as a cultivated state of being that is directly related to sensuous perceptions of the natural environment. We have included Ōtomono Tabito's *tankas* in praise of *sake* among these reflections, although technically *sake* (thought by westerners to be rice wine) is actually a beer. Bacchus, the Greek god of wine, is as courted as the muses by western poets searching for high thoughts, new worlds, divine powers, and, in the words of Anacreon, "the sovereign cure of human woes." In wine the poets arrest their fears of war and pains of lost loves.

Women who would sooner finish off a good Châteauneuf-du-Pape than sip an occasional sherry are sure to find a distinct gender bias in the poetry of wine. It is full of Byronic masculinity and bragging about "Women, Mirth and Laughter" that is all egotism and romance. In the most sensual poetry about wine, seduction and intoxication are metaphors for each other, and since women and wine share the role of adored mistress and source of creativity, they also (fortunately) share the power. John Gay sums this up in a cheerful chorus from *The Beggar's Opera,* insisting that "Women and wine should life employ."

Gay, who identified the inspiration of wine with "Courage, Love and Joy," was also quite as ready to "sing of nappy Ale." That was undoubtedly because he was English, and in his period, the eighteenth century, good English ale and beer had patriotic value—particularly when contrasted with French wine. Anyone who has drunk French beer knows that the English have excellent grounds for claiming the superiority of their barley brews.

Although the earliest Anglo-Saxons had made a kind of honey wine (mead) and the Roman soldiery that settled England had established *tavernae* for wine drinking as early as the first century, it was really the Saxon ale shop that catered to the native taste.*
Ale and beer in England are still associated with pub drinking, where the traditional, dark, room-temperature brews are hand-drawn

---

* For this section we are indebted to Michael Jackson, *The English Pub*, New York: Harper & Row, 1976.

from casks and served in pints that go back to an ancient measure used for corn. When the nineteenth-century rural poet John Clare referred in his "Toper's Rant" to "a fellow/ Who loves to drink ale in a horn," he was alluding to the period of Saxon King Edgar, who had drinking horns marked with pegs to measure socially acceptable gulps. From the custom of horn-drinking contests— revived in John Clare's period—we get the phrase "take him down a peg," which serious English pub drinking is apt to do.

While the brewing of ale is probably as old as the discovery of wild barley, the distinctive quality of English beer derives from the addition of bitter hops to the brewing process; this came about, along with the word *biere,* as a Flemish import of the fifteenth century. Despite some resistance to the addition of hops, the taste was soon acquired and became the hallmark of English beers, as distinctive as the bent roofs of the Kent oasthouses where the hops are dried. "For only just to smell your hops," writes W. H. Davies, "can make me fat and laugh all day." For the Anglo-Welsh poet R. M. Minhinnick, hops are less a laughing matter than a symbol of the bitter lives of the working class, "the bitter that has built the redbrick into the faces of these customers."

Perhaps because of their working-class associations, British beer and ale in their heyday of local brewing were symbols of health and cheer. An anonymous seventeenth-century writer calls ale "immortal," born of Ceres and Neptune, "the only aqua vitae of life." More recent British mythology and advertising declare that "Beer is best," that "Guinness is good for you," and that Mr. XXX (another brand) "raises the standard of living." These claims refer, of course, only to English brews and not to the varieties that were developed from a new strain of yeast brought from Munich to Pilsen in 1842 by the monk who discovered it. The new culture, which causes fermentation at a cooler temperature, spread to Copenhagen, Amsterdam, Dortmund, and Brussels, then crossed oceans to Melbourne and Milwaukee, where brewers began to produce the drink called by the English "lager."

Anyone who has been in the beer halls of Munich and Pilsen can attest that their powerful beer is to be savored most cautiously, though the same can hardly be said of Milwaukee. It's a sad truth that despite the great quantities of beer consumed in

America and the energetic campaigns to romanticize and natural-
ize six packs of every color and dimension, there is "no beer that
will make you a singer"—no beer that tastes remotely genuine.
And so, in "A Small Beer Complaint," we are left to consider the
past and contemplate alternative concoctions like Whiskey Sours,
Bloody Marys, and Long Island Iced Tea.

Among the blithe spirits distilled from grain, fruit, herbs, potatoes, rice, and wine we are free to choose strong poisons of subtle
flavor and walloping potency. Some take them neat, deploring the
novice who contaminates his single malt with water; others know
the fleeting soupçon of vermouth that dignifies gin with the status
of a dry Martini. As the proof gets stronger, writers become plea
bargainers, for the conscience of the serious drinker is a curious
cocktail of glory and guilt, divinity and danger. These are the
swimming sensations that overcome Huysmans's Des Esseintes when
he reaches for the monkish bottle of Benedictine. The world of
spirits is "rich in bliss" for Joseph O'Leary who sings the praise of
Irish whiskey, but it is an "obscure night of the soul" for Ernest
Dowson held in check by the terrors of absinthe. In the literature
of praise and blame, filled with the ironies and paradoxes of *spirit*-
ual allusion, Lord Alcohol is at once the *aqua vitae* and minister
of death.

What is the compulsion to drink? According to Baudelaire, who
recommends perpetual drunkenness on wine, poetry, or virtue, it
is "to avoid that horrible burden of Time. . . ." But as we choose
our poison, so may we choose our motives: the flight from time,
from reason, from poverty, disappointment, responsibility, bore-
dom, or love. "As a result," writes the Zen philosopher Kenkō,
"even dignified men suddenly turn into lunatics and behave idiotically." Sebastian Brant catalogued the diverse lunacies of heroes
felled by drunkenness in his *Ship of Fools*. Yet despite his and thousands of other unflattering, even terrifying, portraits of stupefied
drunks, like Richmond Lattimore's figures "with faces stuffed and
set, and eyes like glass," it has proved as impossible to argue po-
etically as to legislate against drunkenness. This was the lesson of
Prohibition in America, a dry period from 1919 to 1933 that
nevertheless became the "roaring twenties," dominated by the romantic secrecy of the speakeasy. Like most representatives of vir-

tue, the teetotalers who held forth on "the beauties of Temperance" were most tolerable as comic figures and far less attractive than the wicked.

Perhaps self-righteousness was the wrong tack, for without the prejudice of sobriety soft drinks, from lemonade to Coca-Cola, have occasioned diverse praise, and despite all that may be said about caffeine, there has been a positive cult of "calm serenity" mythologically ascribed to coffee and tea. Samuel Johnson, in his 1755 *Dictionary,* defined the sun as "the luminary that makes the day," but for devotees to coffee and tea the day begins, ends, and is measured by fuming cups.

Both were introduced from the East and rose to popularity in western Europe during the Enlightenment. Coffee was a recent invader of England when the seventeenth-century writer Samuel Butler expressed his worst fears that this barbarian "burnt water" and "the vilest of liquors" would drive out chocolate and especially "tea drinks of better quality." This proved not to be the case in England, though chocolate (of South American origins) lost fashion in adult circles. This is quite amusing to Stanley J. Sharpless, who reminds his audience of cocoa's notoriety as an alleged aphrodisiac. Nevertheless, tea became the English drink, requiring stoic patience in the boiling, precision in the steeping, order in the pouring, regulation in the adjustment of milk and sugar, and elegance in the servingware. All very English and totally unlike the grinding, swirling, and gulping of Giuseppi Belli's coffee-bean philosophy! Oddly enough, the hard world of dark coffee has finally touched Japan, the nation so long associated with the aesthetic refinements of the tea ceremony. The new Japan is visible in Hisahi Ito's poem, "Coffee/No Sugar," and also in Japan itself where coffee is everywhere and a cup of perfectly brewed coffee may cost $4.00—the price of political and commercial reality.

Pasteurized, homogenized, and commercialized, no beverage has been more politicized than milk. Its meaning is so fundamentally connected with life that it is the richest potential symbol of life skimmed, condensed, and soured. These and various other adulterations of milk are all to be contrasted with pastoral visions of cows, milkmaids, lapping cats, and suckling babes that answer our nostalgic ideal of virtuous fertility. Pure milk belongs to a world that is pre-war, pre-industrial, and pre-urban.

But the search for milk that tastes of grass and natural cream may be as futile as the search for the fountain of youth. It would take us out of this world. That is where Keats wanted to go when "half in love with easeful Death" he sought a "draught of vintage . . . tasting of Flora and the country green," tasting nothing of "the weariness, the fever, and the fret of reality" ("Ode to a Nightingale").

Facilitating entry into the world beyond has been the traditional employment of sorcerers, witches, and plainer poisoners who are the subject of our final chapter on bewitching brews. Their tonics and elixirs are intense magnifiers of wish fulfillment. They are distillations of human restlessness and longing—for change, for love, power, imagination, and in the end immortality.

How thin the line is between attainment and loss we know from the tragedy of *Romeo and Juliet,* as well as from Circe's hog pen with its baneful wails of pleasure turned to pain. Hang on to your moly root for this last chapter. You'll never appreciate pure water more!

JOAN DIGBY
Oyster Bay, New York

Inspired by Drink

CHAPTER ONE
TANTALUS QUENCHED

Y

JUAN RAMÓN JIMÉNEZ

(S P A N I S H , 1 8 8 1 – 1 9 5 8)

Yes, thirst . . .

Yes, thirst, thirst, horrible thirst!
. . . But . . . leave the water glass
Empty! . . .

—*Translated by Eloïse Roach*

Y

ABRAHAM COWLEY

(E N G L I S H , 1 6 1 8 – 1 6 6 7)

Drinking

The thirsty Earth soaks up the Rain,
And drinks, and gapes for drink again.
The Plants suck in the Earth, and are
With constant drinking fresh and faire.
The Sea it self, which one would think
Should have but little need of Drink,
Drinks ten thousand Rivers up,
So fill'd that they oreflow the Cup.
The busie Sun (and one would guess
By's drunken firy face no less)
Drinks up the Sea, and when he has don,
The Moon and Stars drink up the Sun.
They drink and dance by their own light,

They drink and revel all the night.
Nothing in Nature's sober found,
But an eternal Health goes round.
Fill up the Bowl then, fill it high,
Fill all the glasses there, for why
Should every creature drink but I,
Why, Man of Morals, tell me why?

♟

CLAUDE MCKAY

(AMERICAN, BORN JAMAICA,
1890–1948)

Thirst

My spirit wails for water, water now!
My tongue is aching dry, my throat is hot
For water, fresh rain shaken from a bough,
Or dawn dews heavy in some leafy spot.
My hungry body's burning for a swim
In sunlit water where the air is cool,
As in Trout Valley where upon a limb
The golden finch sings sweetly to the pool.
Oh water, water, when the night is done,
When day steals gray-white through the window-pane,
Clear silver water when I wake, alone,
All impotent of parts, of fevered brain;
Pure water from a forest fountain first,
To wash me, cleanse me, and to quench my thirst!

HOMER

(GREEK, NINTH–EIGHTH? CENTURY B.C.)

From *The Odyssey*, BOOK XI

Next, suffering grievous torments, I beheld
Tantalus; in a pool he stood, his chin
Wash'd by the wave; thirst-parch'd he seem'd, but found
Nought to assuage his thirst; for when he bow'd
His hoary head, ardent to quaff, the flood
Vanish'd absorb'd, and at his feet, adust
The soil appear'd, dried, instant, by the Gods.
Tall trees, fruit-laden, with inflected heads
Stoop'd to him, pears, pomegranates, apples bright,
The luscious fig, and unctuous olive smooth:
Which when with sudden grasp he would have seized,
Winds whirl'd them high into the dusky clouds.

—Translated by William Cowper
(English, 1731–1800)

SIR WALTER BESANT

(ENGLISH, 1836–1901)

A Night with Tantalus

A COLONIAL REMINISCENCE

It was past ten o'clock when the ponies left the hard white road and turned into the dark avenue of palms which formed the approach to the little country box where the two men lived. The night was hot and dry; there was a gentle breeze, but it was the hot wind which lifted the white dust and floated it—all of it, as it seemed—exactly on the level of the riders' breathing apparatus, so as to parch the tongue and dry up the throat.

They were two railway engineers, and they were getting home after a long and fatiguing journey. They had been up and on the line before six in the morning; they had spent the great heat of the day drawing plans in a stifling hot office; they were afield again when the sun got low; they had taken a hasty dinner with the chief, and they were now home again. The monotony of the day, needless to explain, had been varied by many draughts of mingled soda and whiskey.

As they turned into the avenue, one broke the silence, and said briefly, "Whiskey and soda, Jack?"

The other replied, "Two, my boy. It's a thirsty country, but, thank Heaven! there's lashin's to drink."

The tumbledown shanty where they lived had been put up for a hunting-box. It contained one room, roughly furnished with a table, a couple of chairs, a couple of small iron bedsteads, a sideboard, and a safety bin. The box was built of half a dozen uprights, rudely hewn out of trees, and its walls were of thin wood taken from packing-cases. It had a small lean-to by way of veranda. Outside, there was a stable for four horses, a servant's cottage, and a kitchen. Nothing more. Behind it lay a narrow valley running up to the mountains, thick with forest; in front, separated by the avenue of palms, was the long white road; there was no house within five miles. The two men lived here because it was convenient to their section of the line.

They threw themselves off their ponies.

"Arakhan!" shouted one of them.

Now, Arakhan was their groom, cook, and general servant. Nobody else would have Arakhan because he was a convicted burglar, a suspected murderer, and a terrible, black-avised rogue to look at.

"Arakhan!" No reply. "Arakhan, where the devil are you?" No reply.

"Gone a-burgling, I suppose. Got a crib to crack, with a murder. Let's put the ponies in the stable. Hang it! I'm too thirsty to look after them. We'll go and get a drink. Then we'll come back. They won't hurt."

They opened the stable door, led the ponies into their boxes and went out, putting up the bar.

The house door was standing open—it always was open day and night—but there was nothing for any one to steal except the bottles, and they were in the safety bin.

"Phew!" They threw off their hats. "What a night it is! Let's get some drink, for Heaven's sake!"

The speaker drew out a silver box, and struck a light. The match flared up for a moment, and then went out. He struck another. This behaved in the same disappointing manner. "Nasty, cheap, weedy things they are," growled the engineer. He lit a third. "Now then," he said, "where's the lamp?" It ought to have been on the table, but it wasn't.

"There it is, on the sideboard—quick!"

Too late. The third match went out while the lamp was borne from the sideboard to the table.

"Never mind. Here's another."

He lit the fourth match. This burned well and steadily. He lifted the glass of the lamp and ignited the wick. "There!" he said. "Now for the padlock. Oh! give me a soda, quick. I pant—I die."

There stood by the sideboard, screwed into the uprights of the house, a small and very useful article of furniture known as a safety bin. The beauty of this kind of bin is that nobody can take any thing out of it unless he have the secret of the letter padlock which guards the contents. You can see the bottles, but you cannot get them out.

The other man was by this time on his knees before the safety bin. Not praying to the bottles, but using the attitude most convenient to get at the padlock, which was about two feet from the ground, and at his side.

"Hold the lamp, Jack," he said; "I can't see the letters."

Jack took up the lamp. Just then the wick suddenly flared up and went out, leaving a fragrance of oil, but no light.

"What's the matter with the thing?" asked Jack.

"No oil, I believe. The burglar has forgotten the oil."

"Well, we must make a match do. Strike another. I'm like a limekiln."

Jack struck another match.

"Now, then, make haste."

"All right. DROP. That's the word. Here's the D. Here's the R. Confound it!" For the match at this point went out.

"I've lost the letters again. Strike another, Jack. Haven't we got a candle somewhere? Or a bit of paper? Now, then—"

It was pitch dark, otherwise he might have seen his friend turn pale and stagger.

"Make haste, Jack."

"I haven't got any more matches. Give me your box."

The other man rose from his knees and began, carelessly and confidently at first, to search his waistcoat pockets. No matchbox there. He then felt in his trousers pockets. None there. Then he became a little alarmed, and, in some precipitation, began to feel his coat pockets, of which there were many. No matchbox anywhere. He then dragged every thing out. Keys, purse, pocket-book, handkerchief, knife, pencil, foot-rule, pocket-tape, note-book, letters—everything—throwing all on the floor.

"Jack," he said solemnly, after a long search, "are you quite—quite sure that you've got no matches?"

"Quite."

"No more have I. Let's call Arakhan. Perhaps he has come back."

They went out into the veranda and shouted for their retainer. There was no reply; the stars winked at them; they heard their voices echoing from side to side of the narrow valley, growing fainter and fainter.

"He must have another burglary on," said Jack. "The beast is never content."

They returned to the room.

"Hang it!" said the other, "there must be matches somewhere. It's impossible that we should be left without matches. Let's hunt about. You take the table, I'll search the sideboard."

Nothing at all was on the table, except the lamp, which the searcher upset and smashed. The sideboard was covered with a miscellaneous collection of plates and glasses. It was difficult to find anything in such a collection. At the edge stood a large red earthenware jug filled with water. He who looked for matches found the jug, but, unfortunately, found it on the wrong side, so that he toppled it over, and it was broken.

"Well?"

"There are no matches. Try to find the letters by feeling."

"I wish I hadn't broken that jug. Even a drink of water would have been something."

"Well—let us try again."

He found the padlock, and began to feel with his fingers.

"D is a good fat letter," he said. "D. Here's D, I think. Unless it's B. R is—is—I think I've found R. Yes. I'm sure this is R. And here's O—round, fat O. Where's P?" He continued to feel, murmuring hopefully. "Here's P, I believe. Here's P, I'm sure—now then. Hang the thing! The other letters have slewed round."

Everybody knows that with a letter padlock it is necessary to keep the letters in line.

"Try again," said the other man, gasping.

He did try. He tried for half an hour; he tried with patience, and nearly succeeded; then with impatience, and never came near success; while he captured one letter the others slipped round; if he thought he had all, there was one wrong. At last he stood up and wiped his brow in despair.

"Jack," he said, "I should like to curse the thing, but it's no use."

"No use," the other echoed; "I've been thinking the same thing for the last half hour. For such an occasion as this—"

"Look here, Jack. I believe there's a crowbar or a pick in the stable. Let us find it and prize the thing open."

They went out together, and opened the stable door. The ponies occupied two of the boxes. They searched them first. No crowbar there. They then searched the other two, kicking about the litter, and feeling the corners. But no crowbar. Meantime the ponies, finding the door open and no opposition to their going out, did walk out together, and trotted off down the avenue.

"Jack! The ponies are gone!"

They ran out together, calling to the sagacious creatures, who only turned their trot into a run, and, in half a minute, were out in the road and galloping away in the darkness.

"Good Lord! The devil's abroad to-night, I believe."

"They're gone," said Jack. "They'll go off into the forest, and they'll be picked up by a maroon,* and mine was a new saddle. There goes fifty pounds, old man. Because, as for our getting ponies or saddles again——"

"I can't swear, I can't say any thing. I *am* so thirsty."

They crept back to the house, hopeless and crushed. The night was darker than ever; darker and closer, and hotter and stiller. And not a drop of anything to drink—not even cold water. They found themselves once more side by side in front of the safety bin.

"I can feel a bottle," said Jack, with a broken voice. "It's full of whiskey, and the soda bottles are under it."

"I've got a corkscrew in my pocket," said the other. "Who would ever dream of having a corkscrew and no bottle to put it in?"

"The bottle is deliciously cool to touch," said Jack. "It's the only thing that is cool. Can't we cut down the infernal house in order to get it?"

"Look here; tie a handkerchief round your hand, so as to get a good purchase. So. Now, then! foot to foot, hand by hand. Ready? Pull!"

They pulled. They had the strength of ten, because they were so thirsty; the iron bent, but it did not give way, and the padlock held. "Pull again—now!" They pulled like Samson, and with much the same result. Craunch! Craunch! Crush! Crush!

*fugitive slave

They were lying on the floor under a wreck. The uprights of the house had given way with everything—safety bin, sideboard, and the two thirsty men—and all lay on the floor together in mingled wreck.

"Jack! I believe my left thumb's cut off. Are you dead?"

"Very nearly," Jack replied faintly. "There was oil in the broken lamp, and my head's in it."

"Get up and look for the whiskey and the soda. They're somewhere about."

They were. The liquid was on the floor. The bottles were in fragments. It was all over. There was nothing more to be hoped. The worst had happened. Their hands were cut by the broken glass; the side of the house was pulled down; their table and sideboard wrecked; their lamp and their water-jug broken; and their ponies gone. The job was complete. They threw themselves upon their beds and lay there in sleepless silence.

At five in the morning Arakhan appeared. It was beginning to get light, and the wreck was visible. He stood in the door and gazed. Everything broken, and the side of the house gone, and his two masters lying pale and livid on their beds, but not asleep.

"Where the devil were you last night?" asked one of the men from his bed.

"Sahib give leave. Go to port. Yesterday more whiskey come—plenty soda come."

"What?" It was now rapidly getting lighter. The thirsty man sprang to his feet. "Where are they?" Arakhan pointed to the corner of the room. There was the case of whiskey, open. Beside it were soda-water bottles—rows of soda-water bottles—dozens of soda-water bottles.

"And they were here all the time! At our very hands—within reach, and we didn't know it, Jack!"

Gurgle—gurgle—gurgle. It was the opening of the soda. What other reply did he expect?

Y

GILLIAN CLARKE

(WELSH, 1937-)

The Water-Diviner

His fingers tell water like prayer.
He hears its voice in the silence
through fifty feet of rock
on an afternoon dumb with drought.

Under an old tin bath, a stone,
an upturned can, his copper pipe
glints with discovery. We dip our hose
deep into the dark, sucking its dryness,

till suddenly the water answers,
not the little sound we know,
but a thorough bass too deep
for the naked ear, shouts through the hose

a word we could not say, or spell, or remember,
something like "dŵr . . . dŵr."

♈

ASHER BARASH

(ISRAELI, 1889–1952)

Hai's Well

I

In a poor little colony which lies high up in Lower Galilee there was a little man who had a little wooden hut, a dunam of land, and a wife and three children. The man went by the name of Nathan Hai. He was a farm-worker who had gone through all kinds of transformations in Judah and Shomron and Upper and Lower Galilee. He had fevered a lot and affectionately cursed the ways of the old and new Yishuv alike, besides using a sharp bachelor's tongue to tease the girl workers and mock at marriage. But, when he was thirty-five years old and looked like a dwarfed and wrinkled olive trunk, he took to wife the "Rosh Pinah seamstress," who was younger than he, and settled down in the little colony. Within five years they brought three healthy sons with good appetites into the world, and the burden of life rested thenceforward on the two of them.

How did he come to the name Hai? After all, he was not one of the Sephardim* among whom that name is common, but a fellow from the neighborhood of Dubno. The truth was that the name "Hai" was short for "Hai vekayam" (hale and hearty). For that was his regular reply whenever anybody asked him how he was. "How are you, Nathan?" "Hai vekayam!" And to be even more precise it should be added that even this expanded name of "Hai vekayam" was only a translation made by one of his friends, who was hot for the Hebrew language and translated into it the Yiddish reply that he had been giving for several years, namely "Hai-gelebt!" (Hai-alive). Nathan cheerfully accepted the translation, but as time went on and he said less because his troubles

---

*The occidental branch of European Jews who settled in Spain and Portugal, Greece and England.

were growing greater, he would simply answer, "Hai!" And so the short name remained and everybody from Dan to Beersheba knew him as Nathan Hai.

Like a dozen other fellows of his own type in the Second Aliya, he was a kind of monomaniac. And what was his particular mania? Water! The redemption of the land and hence, obviously, the redemption of the nation and the Ingathering of the Exiles depended only on water. If Eretz Israel* could only get enough water it would be a paradise. And he had bundles of proof, both by word of mouth and in writing, from hundreds of sources. He would begin with a verse at the very beginning of Genesis, "And a river went forth from Eden to water the garden" (which goes to prove that if there had been no irrigation, there would not have been any Garden of Eden), and go right up to the famous phrase of Kaiser Wilhelm to Herzl† as he sat on horseback in the heat of the day near Mikveh Israel: "It needs water, plenty of water!" And then he would add his own experience.

For Nathan Hai did not rest content with talking about water but was, so to say, the servant of water everywhere. He investigated the water situation in the country from the salt water in the sea and the exceedingly salt water of the Dead Sea to the waters of the rivers and the brooks and the wadis; spring water and flood water, underground water and rain water, upper water and lower water, water for irrigation and water for power. For months and years he took a hand with the well-borers in various parts of the country. For months and years he toiled in the motor cabins of the orange groves, working at draining and drying swamps and laying irrigation pipes. In brief, wherever there was water, there he was to be found. And they say that the old worker and writer who was the teacher of all the workers slapped him on the shoulder once and, speaking of him, quoted a verse from the Book of Job: "He gives blossom from the scent of water."

And now came something queer. This Nathan Hai, the water man, chose as his dwelling place a desolate colony in Lower Galilee, which was white with dust the greater part of the year, in order to raise his hut there and settle down. One might have sup-

---

*Land of Israel; the reference below to Mikveh Israel alludes to a ritual cleansing bath.
†The Austrian emperor said this to Theodore Herzl, the founder of Zionism.

posed that in this way he wished to symbolize his great longing for water. For, if he had dwelt in one of the Sharon colonies where there is an ample water supply, the source of his longing would have dried up. But in the Galilean colony, swooning with thirst, the staff of his longings put forth blossoms like Aaron's rod in the Bible; and he never wearied of talking about water.

In the center of this colony there was a well with a wheel over it, whose water came from the rains in their season. If the winter was rainy, the well was full and it had enough water (only for man and beast and the few yards of green round the houses) until the beginning of August, after which nothing was to be brought up from the well bottom save mud and mire. If there was little rain in any year the water lasted until mid-June or, at most, to the beginning of July. After that there was work for the water carriers (and naturally Nathan spent no small amount of time on that job as well). Morning and evening they had to take the mule and the big barrel on two wheels and fetch water from the distant fountain on the Arab land, water that had to be bought for good money. If the water carrier came home late in the evening there would be a cloud on the faces of the colony people, as they asked themselves what they and their beasts would have to drink.

But Nathan was not the fellow to see folk suffering and do nothing. From the time he became an established citizen among all the others in the place (by right of his cabin and his dunam of course) he would get into touch with every "factor of consequence in the Yishuv" in order to raise the question of water for the Galilee settlements in general and, in particular, his own colony which suffered from the shortage more than them all. It cannot be said that Nathan Hai's efforts did not bear fruit. He set all the wheels in the Yishuv moving about this business of water, from the staff of the Baron de Rothschild in Palestine and abroad to the Jewish National Fund and all the departments of the Jewish Agency. He kept them all busy with letters and memorandums and interviews about the water without which the colony could not live. No excuses helped. They came, they investigated, they sent experts, they hewed and they bored, and they bored and they hewed, inside the colony, down on the hill-slope on the one side and then down on the other hill-slope; a mile away from the house that lay farthest north and a mile away from the house that lay farthest south. They bored and they stopped, and they went back and they bored again. Drills were broken, workers were injured, and one who fell from the stand broke his spine and was crippled for life. But they did not find any water. In one spot they bored for six weeks and they found moist earth, and the pump even brought up a few pails full of fluid mud. Some of the local inhabitants began dancing an Arab "Debke" all round; but after they bored a little more it all stopped and once again the drill brought up only dry cold unfriendly gravel, to the disgust of the inhabitants.

But Nathan never ceased prophesying: "There is plenty of water in our good earth. It has an artery throbbing like the heart's artery in the human body. It is only necessary to find the pulse, to put up the drill at the right spot."

And one day (nobody quite knew how he managed it) Nathan fetched out of the Negev an Englishman with a little yellow beard, wearing a Bedouin *Kefiya* and *agal* round his head. This fellow had spent many years in the Sudan, spoke Arabic with an English accent and a lot of hard guggings, and he carried a little wand with which he walked about like a lizard, touching the surface of the ground, holding the wand and watching it trembling, and deciding accordingly whether there was water at that spot or not.

The Englishman decided on a new spot for boring a well. But after all the bitter experience and waste of money there was no longer anybody prepared to invest the first hundred pounds that were needed to begin the work. The local inhabitants fed the Englishman on the best to be found in the house of the *Mukhtar*, and also presented him with five pounds that they collected in the colony before giving him an honorable send-off. But no boring followed.

Meanwhile Nathan Hai lost his strength. He was already forty-three and four and five. His hair, which had been curly in the old days when the Rosh Pinah seamstress had stroked it with trembling fingers and whispered, "Nathan, you have nice hair"—most of that hair had fallen out, leaving a sunburnt scalp that was as smooth and gleaming as silk. The three children, all of them boys, did not make the little hut any quieter, particularly as she, Eva Leah, always had to be sewing with a machine in order to make most of the living. She used to sew for the Arab women of the surrounding villages too; and he, fully aware though he was of the husband's duties to his wife and those of a father to his children, earned very little. Sometimes he would have to go far away for weeks on end in order to look for work; and during his absence his dunam would suffer from jaundice and baldness.

Not that he ever became melancholy. That is, he looked miserable enough, particularly his eyes which had sunk deep and burned with repressed unhappiness. But his mouth still knew its job. He would make brief jokes, and when he came to talk about water it seemed as though his own trunk had been watered and the living water had entered his veins. His eyes would gleam and his tongue would serve his flights of imagination in lively fashion. "Before long any amount of water will come up to irrigate the land of the colony and make it yield. It is pouring along under our feet. Can't you hear it? Just listen carefully and you will!"

The listener would slap him on his shoulder, look him in the eye and ask:

"And what's the news with you, Nathan, about making a living?"

"We live, so-so—Hai vekayam. Hai!"

II

When their youngest child was six years old, after an interval of six years that is, it came about that Eva Leah found herself in the family way again, for the fourth time. The whole household became apprehensive. Nathan Hai himself also grew afraid. That was all that was missing! As it was the three they already had were wandering about like starving jackals, in rags and tatters. Eva Leah sewed for everybody, yet for her own children she could not sew shirts or a pair of linen trousers. She simply did not have the time. And if she were to take the time off to sew for them, what would she have to put in their mouths? And now this was coming! After they thought they were finished with any more children!

The local nurse and midwife scolded Eva Leah, abused her thoroughly and demanded and insisted that she should go to town and do what had to be done. "Human beings aren't swine!" she permitted herself to say coarsely. She was an old maid with principles. (Once upon a time she had been one of those who had "gone to the people in Russia.") For a moment Eva Leah hesitated saying to herself: "Maybe she is right and it would be worth doing what she suggests." But when she said as much to Nathan he opened a pair of startled eyes at her, then spat for all he was worth and cried in a voice which was not his own:

"Listen to a block of dry wood like that? No, as long as I am Nathan Hai, you won't do anything so abominable!"

And the item was struck off the agenda.

It must be confessed that from that day forward there was a different mood in Nathan. He seemed to be transformed into a kind of tense machine on springs. He was on the move all the time, travelling to Tiberias or Haifa, to Afulah and Nazareth, even to Tel-Aviv and Jerusalem. He would spend a few days or weeks there, then came home fetching in his rucksack a few things that they required. Maybe some tinned food or some cloth or knitting materials and so on. And his tired-looking purse would also have a little ready cash.

So there was a little light in the hut. Nathan's face looked worse, to be sure. He grew thinner and seemed to shrink. The silken bald patch turned to a burnished copper, and his eyes flamed as though he had the fever. But he was in a good mood. Often he would answer those who asked how he was with his old phrase of, "Hai-gelebt!" but would at once correct himself: "Hai vekayam!" The children were dressed in more orderly fashion, for if he spent a few days or weeks at home he compelled Eva Leah to turn down a few customers in order to sew some clothes for them.

The day of the birth was on the way. They reckoned that the child ought to be born in the middle of August.

The ninth of Ab passed at the beginning of August and Eva Leah completed her full term. Everything was ready for the birth. They had already spoken to the midwife. Her fury had not quite died down at the "barbarism," but she had no choice save to accept the fact, and she came into the hut from time to time to see that everything was in order.

On the eighteenth of Ab the birth-pangs came. Nathan quickly filled two tubs of water. Although Eva Leah had not cried out, the nurse heard and came running, only to see almost at once that the birth was not in order. After an hour of sweat and toil she realized that the doctor had to be brought from the neighboring colony because the danger was increasing. She told Nathan so. He asked no questions, grabbed his stick and dashed off to the colony.

It was the forenoon, in the heat of the day. High up in the heavens sailed distant white wisps of cloud, enjoying a sun bath. The little colony was silent with its poor little houses. The few gardens were grey with the remains of scorched vegetables, looking like stains of rust on zinc. Only the pruned eucalyptuses at the wayside rose green with their young branches. The way down the slope to the wadi was thick with beaten dust. In the skies three vultures were circling in a triangle: One in the north, one in the west and one in the east. Nathan noticed them. It immediately occurred to him that they were coming from the water, one from the Kinnereth,* one from the sea and one from the Jordan. This thought gave wings to his feet, so that he did not notice the sweat running down like a fountain of water from his head to his collar. It took an hour through the wadi to reach the big colony, but he ought to get there in half an hour, for he was running. Why had he not taken a horse or donkey? It simply had not occurred to him. Now it was not worth his while going back. He had already passed a quarter of the way.

Nathan passed the cemetery on the little hill. He glanced at the handful of scattered tombstones, two of which stuck out so importantly while the rest lay like stones in the field. He suddenly felt afraid at the sight of the cemetery, and the thought of Eva Leah twisting and turning in her birth-pangs, all in ever-greater danger. He went even faster down the slope, running with his stick ahead of him. The sweat poured from his head over his face, into his mouth, and down his neck and over the hair of his open chest. When he got to the bottom it was no longer so hot and he felt a little easier. He removed his hat, held it in his other hand and allowed his gleaming bald patch to absorb the sun.

All of a sudden at the entrance to the wadi, he felt a kind of slight stab in the head. His legs began to quiver. They seemed to grow light, lifted themselves a little from the earth and fluttered

*Lake Tiberias.

in the air. His heart beat as though it was running away with itself. The light turned suddenly dark. He fell on his stick, quivered a little, turned over at the wayside and slipped down into a shallow ditch.

III

Two hours and more passed but Nathan did not return. Neither did the doctor arrive. The midwife came out of the hut with her hair in disarray, and shouted in a way that the little colony had never heard since it was founded. Nathan's three children immediately stopped their play, startled; they sprang up and stood staring as though they were senseless. One by one came grownups from the houses. The woman shouted at them:

"What are you standing like blocks of wood for? Go and fetch the doctor quick! Eva Leah is dying!"

Within half an hour one of the local lads had fetched the doctor on his cart. But the doctor found Eva Leah lifeless. And the child remained within its mother.

A village Arab found the body of Nathan Hai in the ditch five days later. He saw kites busy and gathered round there. He approached and recognized the dead man. So he went and told the *Mukhtar*. When they brought him away from there his flesh was already going. His face and eyes had been pecked by the beaks of the birds of heaven.

They buried him as he was, without cleansing him, beside Eva Leah's grave. The three orphans were shared out to three houses, one to each. The little one was adopted by the midwife, who had fought like a lioness against death but could not prevail.

The colony mourned grievously. In one day three souls had been cut off and a family had been uprooted in the little community. Who can understand the cutter-off of life?

IV

Once Nathan Hai had been brought to burial after the fashion of Israel he was eased of the burden of his life, which had been beyond his strength. Eva Leah lay not far from him, her child within her, and rested forever. "There rest the exhausted of strength" and who was so exhausted of strength as to compare with Eva Leah?

Even before her marriage, as long as she had been in Eretz Israel, she had been harnessed to the yoke of hard work; first as a worker, then as a domestic help, finally as a seamstress. Even after she had married Nathan and given birth to three sons one after the other, each coming before the other was big enough to look after himself, she had continued to sit bent over the sewing-machine which had sucked up the rest of her blood. But she was a good person of spirit and character, and no matter how she suffered she was never heard to make a complaint. Now she had gained the rest she deserved. The children had been orphaned of their father and mother, but were better off than they had been. Each of them was in a good home, his bread provided at the proper time and his clothes not lacking. Best of all was the little one. For the captious and bitter old maid gave him all her pity and love. He was small and tender in her eyes.

And he, Nathan Hai, had attained full rest. His bones, worn and broken with hard work, now took it easy. His brain, which had grown weary with concern and alarm for the souls depending on him, could also rest and devote itself entirely to the one thought which had filled him from head to foot while he was alive; the thought of water.

For days and days, for months and months, he lay listening, sending forth his will, one might say, like delicate antenna deep into the earth. He forgot everything. He forgot himself and the whole world. Somewhere, in some hidden place a pulse of water was babbling and throbbing, longing to be revealed. Summer passed, winter vanished, a new summer came, and his will grew tense as a violin string, while his water sense grew keener than ever.

The following summer, in the heat of the month of August, a year after he had fallen at the wayside, a faint sense of moisture reached him. First he did not know from which side it came. He grew very excited. He had to gather all the strength of his will in order to sense precisely where it came from. At length he realized it exactly. It came to him, a cold and pleasant stream, from one specific point. There it was! The vein of water gathered in his awareness. It was not far away. He had had to labor with his awareness for a whole year until he found the spot. Ha ha, it was certainly not the spot to which the English water diviner had pointed! No, the fountain was here, below the cemetery. Exactly ninety yards in a straight line west of his grave! Ninety yards, according to the numerological value of *mayim*, the Hebrew word for water!

All his bones rejoiced. Now he knew what he had to do. He knew. A dream. The dream he had dreamed on earth had come about. Now he knew what he had to do. The locked and sealed water must rise up.

V

The following summer, during the hottest days of August, Nathan Hai appeared in a dream to the *Mukhtar* of the colony for three nights running. He appeared with radiant face, and this was what he said: "Wake up and get up! Go out and bore! Exactly ninety yards in a straight line west of my grave! Thirty-three yards down the water is waiting for you. Don't delay! For a whole year I have been bringing the water up to that point. Don't miss the chance. I shan't be able to keep it there very long. Uncover it!"

On the first night the *Mukhtar* woke up and in the darkness told his wife about the strange dream. They both decided it was nonsense. But next day the man went about all day long not knowing what to do. The dream would not leave him alone even for a moment. After the second night, when the dream was repeated again, he told it to several of the local people, and one of them went so far as to remark: "It isn't just nonsense." After the third night the *Mukhtar* summoned a meeting of the committee and they decided to try. Thirty-three yards—the cost was next to nothing.

Now it so happened that not far away the water company was boring a well. They had already gone down more than two hundred meters. They had already been boring for three months. They had already made their way through two strata of rock, and now they had reached a third which was even harder. But there was no sign of water. Work had stopped three days before. The borers had simply grown tired and given up. Their tools lay where they were, like dead corpses.

Two committee members went to the company in Haifa and deposited thirty-three pounds, according to the number of yards that they wished to have bored. Next day the well-borers came to the spot, which was already marked with an iron spike. They dug a little, put up the stand for the drill and began to drill.

Lots of jokes were heard, as they worked, and the deeper the drill bit, the lower grew the spirits of the people and the smaller their faith in the *Mukhtar*'s dream. After the drill had passed the thirty-yard mark they all but gave up, and felt that they wished to ask the workmen to stop; but were ashamed to do so.

And then, all of a sudden—

It was noon, the heat of the day, just the time at which Nathan Hai had run off to summon the doctor and had fallen while running and dropped into the ditch. White clouds were sunning themselves in the sky. Three vultures were circling in a triangle. The final thuds of the drill sounded in the silence of the colony— and the iron rod in the hands of the drillers suddenly beat, while a lapping sounded from the deeps. Before they could see what had happened a sound suddenly rolled and echoed all round. Water! Water! The little pump standing there suddenly emitted a jet of water, pure water from the hole. It gleamed like crystal in the sun, fell to earth and melodiously flooded all the neighborhood. And the pump went on pumping the precious fluid on the ground.

There was not a living soul in the colony who did not come to see the sight. All of them stood over the little pool, gazed through tears and trembled with excitement and joy.

The experts measured the force of current. There were two hundred cubic meters an hour. A fountain of salvation had been opened to plants and living creatures. Now the colony would begin to grow and flourish. The words of Nathan had come about.

VI

Before long a big reservoir was put up. It was big and tall and stood on five tremendous pillars, four at the corners and one in the middle. A proper pumphouse was also built. And the village water festival was held.

Many people came from settlements close at hand and far away to rejoice with the people of the colony, whose thirst had been quenched with much water. Each one brought his gifts of all wherewith the Lord had blessed him. Round the well they planted saplings and fresh flowers that were amply watered. The whole of the square intended for the rejoicing crowd had been besprinkled with water. Every white shirt, every white blouse, whether of man, woman or child, was adorned with a green twig. Neighbouring Arab horsemen took their places amid the horse-riders of the colonies. Tables on trestles stood ready, loaded with good things to eat and drink.

Three choirs of children from three schools in three colonies stood on the platform, the music teacher keeping them quiet with his conductor's baton. They were waiting to sing the song "And you shall draw water with gladness from the fountains of salvation."

In the front row, like pioneers before the choirs, stood the three children of Nathan Hai, each smaller than the next, all dressed in suits that were as white as snow.

In front of the pumphouse, over the iron door, was an inscription in large letters:

HAI'S WELL
This is the well of Nathan Hai, and such is its name and fame forever.

But one man, the man who writes this, looked sorrowfully at what he had added, while nobody had noticed, in chalk on the iron door in small printed characters: "Let Eva Leah also be well remembered."

—*Translated by I. M. Lask*

Y

JONES VERY

(AMERICAN, 1813–1880)

Jacob's Well

Thou pray'st not, save when in thy soul thou pray'st;
Disrobing of thyself to clothe the poor;
The words thy lips shall utter then, thou say'st;

They are as marble, and they shall endure;
Pray always; for on prayer the hungry feed;
Its sound is hidden music to the soul,
From low desires the rising strains shall lead,
And willing captives own thy just control;
Draw not too often on the gushing spring,
But rather let its own o'erflowings tell,
Where the cool waters rise, and thither bring
Those who more gladly then will hail the well;
When gushing from within new streams like thine,
Shall bid them ever drink and own its source divine.

Y

HENRY DAVID THOREAU

(AMERICAN, 1817–1862)

Such Water Do the Gods Distil

Such water do the gods distil,
And pour down every hill
  For their New England men;
A draught of this wild nectar bring,
And I'll not taste the spring
  Of Helicon* again.

*The sacred fountain of the muses, called the thespian spring, that was said to issue from this Greek mountain.

Y

ROBERT BURTON

(E N G L I S H , 1 5 7 7 – 1 6 4 0)

From *The Anatomy of Melancholy*

Pure, thin, light water by all means use, of good smell and taste, like to the air in sight, such as is soon hot, soon cold, and which Hippocrates* so much approves, if at least it may be had. Rain water is purest, so that it fall not down in great drops, and be used forthwith, for it quickly putrefies. Next to it fountain water that riseth in the east, and runneth eastward, from a quick running spring, from flinty, chalky, gravelly grounds: and the longer a river runneth, it is commonly the purest, though many springs do yield the best water at their fountains. The waters in hotter countries, as in Turkey, Persia, India, within the tropics, are frequently purer than ours in the north, more subtile, thin, and lighter, as our merchants observe, by four ounces in a pound, pleasanter to drink, as good as our beer, and some of them, as Choaspes in Persia, preferred by the Persian kings before wine itself.

Clitorio quicunque sitim de fonte levarit
Vina fugit gaudetque meris abstemius undis.

* Ancient Greek physician.

[Who once hath tasted the Clitorian* rill.
Spurns wine, and of pure water drinks his fill.]

♟

MARVIN COHEN

(AMERICAN, 1931 –)

First Came Water

What a waste, that water should be so wet! I think wetness is a superfluous property, that makes water unnecessarily redundant. Therefore water should sacrifice wetness (being wet enough), and donate its essentially surplus excess of property to painfully dry things, like barren rocks, thirsty plants, desolate deserts, and decaying wood. But this would be socialism, and would horrify the capital out of my American soul. So I guess inequality has its place, if free enterprise is to endure, and so let water maintain its monotony—I mean monopoly—on liquid, fluid wetness—or even on still wetness, in a stagnant pool or quiet ocean, if it so chooses. *Laissez faire* is my motto, and if water owns a vast wealth of wet— has invested its soul in wet, and floated liquid stocks in it, and commands a bond with it, so that the quality is inexhaustible by quantity—then let it be. If survival makes the fittest, then, by the law of the jungle—but preferably that of the ocean—the rivers and lakes of the world ought to acknowledge water. And when it rains, that's still further evidence. What is one to do?—defy God? No, deify Him, or we'll be all wet. So if my theory holds water, and slakes the thirst of truth, then no use crying vain tears (artificial water, and not the real thing itself, but a salted man-made product, synthetically caused by grief), but let's base our life on water, and all praise it, the good basic element that rounds out the addition of our life, a fertile force opposed to sterility; let's join in

---

*Water from the fountain in Arcadia claimed by Ovid to destroy the desire for wine.

the swim, laud our origin, unite, and say, "God bless water!" Once having said that, we're free to keep out of the rain: *if* there's rain, which is not always guaranteed, especially when the weather is good.

Good old water. However, I prefer alcohol. Only the dead can get drunk on water, if they've drowned in it. I prefer to keep my feet dry, and my tongue wet, and intoxicate my inward spirits with an outward addiction to the beverages of fire—whisky, for instance. Without water, thank you. Perhaps with ice. But here's hoping that the ice doesn't melt, not until emptiness has removed the contents of my glass, so that staggering I yell: "Holy water! Baptize me in it!" and collapse, a happy man, solid, formed of earth. Hurray. With tons of water in me, the fire of my spirits tinted and tainted and taunted by the universal life-giver. Three cheers for water. When I stand up, I'll know what I'm saying, but meanwhile, I'm drunk on life, and drowning in all that blesses. What does air mean, when water suffices? Again, I'm a fish, and reverse evolution. I'll swim back to the paramoeba, and declare my first cell, and be pre-born. With all of destiny before me, man not even conceived, and God a beardless youth, surviving His ad-olescence, creating the future with some clay, to which He sprin-kles the addition of water, to modify, perhaps to purify, as the unformed monsters clash in His brain, the creatures fertilized to earth, born of the spray. Ah, the sands of time, the beach of spawned activity, and always the tides moving in. Thank you, God. I have enough water. The water is fine. That's quite enough. No more, thank you. Now it's too much, and my poor breath is gone. Down I'm under, in this kingdom of wet, my soul dry. And now, only forever remains. Dry, but preserved. No more liquid fury, the shifting shapes. Eternity's monument, out of the sea. Where no ship dwells, no fins or gills exist, but simply, without time, some-thing dry and breathless, a deep container, filled always with emp-tiness, that spills over. Beyond the brim, an everlasting fountain. Or a well, holding to the middle of the earth, from which a great nothing is drawn, pumped, yielded, a glory renewed in its birth, always but never the same, to which water and earth were doubly joined, born to be dead simultaneously. What a unity. Where wet and dry, death and life, nothing and all, can't be separate, and have their great unification. Then what can divide us? How can we be alone? All this, everything, and nothing. And nothing being so abundant, it contains everything. Surely, what more is needed? Water can slench the thirst, anything can do anything, and great Nothing contains all. The sea, let us go to the everlasting sea. And see our simple beginning.

ROBERT HERRICK

(E N G L I S H , 1 5 9 1 – 1 6 7 4)

To the Water Nymphs, Drinking at the Fountain

Reach with your whiter hands to me
   Some crystal of the spring,
And I about the cup shall see
   Fresh lilies flourishing.

Or else, sweet nymphs, do you but this,
   To th' glass your lips incline;
And I shall see by that one kiss
   The water turned to wine.

CHAPTER TWO
THE SOUL OF WINE

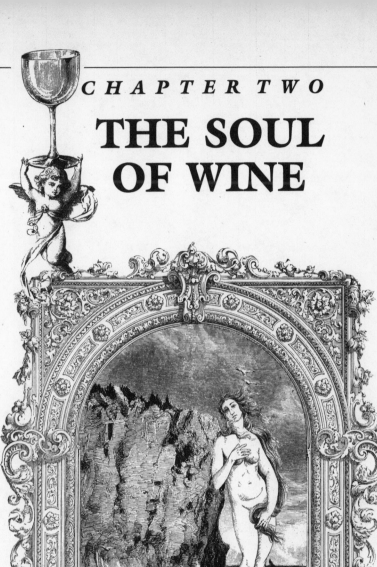

♀

JOHN GAY

(ENGLISH, 1685-1732)

From *The Beggar's Opera,* Act III

FILL EV'RY GLASS, &C.

Fill ev'ry Glass, for Wine inspires us,
  And fires us
With Courage, Love and Joy.
Women and Wine should Life employ.
Is there ought else on Earth desirous?
  Fill ev'ry Glass, &c.

♀

ANONYMOUS

(AMERICAN, TWENTIETH
CENTURY)

An Old Southern Story

A revenue officer caught a bootlegger with a lot of jugs in his truck and asked what was in them.

"Water," said the bootlegger.

The officer didn't believe him and opened one of the jugs to taste it.

"Looks like wine," he said, "and tastes exactly like wine."

"Lord, Lord," cried the bootlegger. "Jesus has done it again!"

JEAN LE HOUX

(FRENCH, 1551–1616)

Four Poems

1

With my back to the fire, elbows on the table,
Among wine flagons I feel happy and stable.
 Just like a newly hatched chick
I will not allow myself to die of the pip
When the face chokes crimson in a strangler's grip
 And the nose glows violet and sick.

Now when my nose turns aquamarine or shines red
It wears the colors my mistress likes on my head.
 Wine brings a flush that can alter
A sickly complexion in pale, puny cheeks,
Not like having the tincture of two sallow leeks
 As a timid drinker of water.

I'm all against water—at least for drinking
In case of hydropsy as I'm thinking
 And I might perish if I drink it.
Water has no bouquet so why accept a beverage
As my wise neighbour says that comes from no vintage.
 Water I like not a bit.

Whoever loves wine is wise and in perfect good health.
The dead can no longer drink it even in stealth!
 Who knows if I'll be
Alive and kicking tomorrow. O sing melancholy,
I'm going to drink with this company so jolly.
 Follow all those who love me.

2

Beautiful nose as rich as rubies now sing
 Of white wine with a delicate blush
Also of wine enriched and fit for a King,
 Deep red that sparkles with a purple flush.

Great nose! Who holds you back from the beaker!
 You who can judge a bouquet and never falter.
You no longer have the nose of a sorry drinker
 Who, poor soul, drinks only insipid water.

That Indian cock has a bright throat like you.
 Now tell me how many of the richer
Folk has such a nose painted with a glowing hue.
 Very few I'm afraid who swill from a pitcher.

The glass is a paintbrush that gives you highlights
 And sparkling wine is the color
That paints you, nose, like a moon on cloudy nights
 Adding luster to your pallor.

Wine curtains the eyes, but are they the Princes Elect?
To cure all ills give yourself a good souse
In delicate vintages so rare and select.
I'd rather lose two windows than a whole house.

3

Nightingale, musician, Queen of the air,
Appearing in Spring's flowering so fair
 With your trilling song floating across the lawn.
But nightingale if I were a bird like you
With flowing wine I would salute the dawn,
 Singing sweeter than any of your crew.

The truth about me I'm sorry to say
I'm rather inclined to sleep late in the day,
 And wouldn't stir at such an early hour to sing

But having snatched a little beauty sleep
And doing boring things like counting sheep
   I would whet my crop with wine and ring.

To wake up as early as you every day from bed
I think my spirits would be as heavy as lead.
   But if I could drink wine, a quality label,
(For that's the best way to rouse me from the dead,
As there's nothing like drinking for getting ahead)
   I would soon be on my feet steady and stable!

4

Bacchic Comparison

Myself without a glass and drink
Is like a snail without its shell,
A chaffinch without its pink,
A church without its bell.

It's a hunter without his horn,
A German soldier without his boots,
A princeling with nothing to pawn,
A shepherd without his flutes.

It's an old soldier without his plume,
It's a carpenter without his knife,
It's a drummer without his fife,
It's a pessimist without his gloom.

Without wine I lose my spirit:
It suits me as fine to drink wine
As a lawyer issuing a writ
To quench my thirst from the vine.

I have come to do battle
In the name of love with my friend,
The wine in this bottle
I'll drink to the end.

—Translated by John Digby

♆

FRANÇOIS RABELAIS

(F R E N C H , 1 4 8 3–9 5 ? – 1 5 5 3)

From *Gargantua and Pantagruel*, I, 5

PALAVER OF THE POTULENT

Then they returned to Grangousier's. Straightway the flagons danced a jig, hams were trotted out, goblets began to fly, carafes to caper and glasses to tinkle.

"Draw my wine, boy!"

"Give me *my* glass!"

"Fill mine up!"

"Water in mine, please."

"I want mine without. So, friend!"

"Polish off that glass, lad: drink up; look sharp now!"

"Produce the evidence: *vin rosé,* bailiff, and let the glass weep from the fullness of its heart!"

"A truce to thirst! Wine is victorious!"

"Ha, shall I ever be rid of the false fever that parches my throat?"

"Upon my word, gammer my darling, I can't feel my way into drinking!"

"You've taken cold, my sweet."

"Ay, so I have, I'm thinking!"

"By the belly of Bacchus! Let's talk about drinking!"

"I only drink at my hours; capricious, I am, like the Pope's mule."

"Pass me that flask shaped like a breviary; I'll drink no other wine; a pious man, I; I'll have breviary wine, ay, like a f-f-fine F-f-franciscan F-f-father S-s-s-superior."

"Which came first: thirst or drinking?"

"Thirst came first. Who would have thought of drinking without being thirsty, in the age of innocence?"

"Drink came first, for *privatio presupponit habitum,* says the law:

privation of something presupposes being accustomed to it. A clerk, I. *'Faecundi calices quem non fecere disertum'* Horace put it aptly: the cups of the talkative, but not the eloquent!"

"The age of innocence, you said? We poor innocents drink all too much without being thirsty!"

"You innocents drink without being thirsty? Well, I'm only a miserable sinner, but I'd not drink unless I were dry. Pah! if I'm not thirsty when I drink, I forestall my thirst: I drink to slake the thirst to come! I drink everlastingly: an eternity of drink, a drinking of eternity!"

"A song! A drink! Let us sing a religious chorale!"

"Who will corral wine into my glass!"

"Ho! I'm drinking by proxy it seems; my glass is empty!"

"Do you wet our whistle for the pleasure of being dry later or do you let yourself get parched for the pleasure of wetting your whistle?"

"Bah! I know nothing of theory but when it comes to practice, I'm not so badly off."

"Make haste, boy! My glass!"

"I wet, I dampen, I moisten, I humect my gullet, I drink—and all for fear of dying of aridity!"

"Drink up, you'll never die!"

"Unless I drink, I desiccate; desiccation means death. My soul will fly to seek the moisture of a frog pond, for the soul, says St. Augustine, cannot dwell in a dry place."

"O butlers, O wine stewards, creators of new substantial forms, make of the abstainer that I am a nonpareil drinker!"

"May I be perennially macerated in wine, through these, my sear and sinewy guts!"

"He drinks in vain who does not enjoy it!"

"This potion runs through my veins; my pissing tool shall have none of it!"

"I'd gladly wash the tripe of the calf I dressed this very morning. What calf? My belly. Dressed how? For the table!"

"I have ballasted my stomach choke-full!"

"If the parchment of my writs drank as avidly as I do, no trace of writing would remain upon them. Then, sirs, what bitter wine my creditors would taste when called upon to produce the evidence of my debts!"

"It's your hand makes your nose red . . . you raise it too often to your lips!"

"Ah! how many more beakers shall enter here, ere this one flow out of me!"

"You'll scrape your snout against the bottom if you drink so shallow."

"What are those flagons there for? Decoy birds, *I* say!"

"What is the difference between a bottle and a flagon?"

"A vast difference. The bottle's a spirited man who'll need but light corking to keep him from spilling. The flagon's a spigotted woman who'll need hard screwing to keep her from spitting!"

"Ha! Well said, friend!"

"Our fathers drank lustily and emptied their pots!"

"Well hummed, well bummed: a fine movement of music and bowels. Now let us drink up!"

"This round will wash your guts for you! Have you anything to send to the river? That's where tripe is washed: give this drink your message!"

"I drink no more than a sponge!"

"I drink like a Knight Templar."

"*Spongia*, a sponge, eh? Well, I drink *tamquam sponsus,* like a bridegroom."

"And I *sicut terra sine aqua,* like earth without water."

"Give me a synonym of the word 'ham'!"

"A subpoena served upon thirst; a compulsory instrument in the jurisprudence of drinking. A pulley, too: you use a pulley to get your wine down into the cellar, and ham to get it down into your belly."

"Ho, lad, come here, come here! Another drink! I've not my full complement of cargo! *Respice personam,* respect the person; *pone pro duos,* pour for two! Your marvel that I say *duos for duobus,* that I, an erudite man, make such an error? Let me tell you that *bus* is obsolete. *Je bus* means I *have* drunk, in the *past;* while our drinking is forever conjugated in the present tense."

"If I could up my limbs as well as I down my liquor, I'd be pacing the ceiling long ago."

> *Drinking made our magnates wealthy.*
> *Drinking makes our bodies healthy!*

Thus Bacchus won the giant realm of Ind!
*Thus Vasco found that mine of gold, Melinde!**

"A touch of rain allays a lot of wind; long tippling breaks the thunder."

"Would you suck my left ballock if it piddled out such liquor?"

"I always hold it after drinking!"

"Here, boy, a drink, ho! I herewith submit my credentials, certificated by process of law, as a candidate for your favors!"

"Drink up, Will, there's a mugful still!"

"I herewith file a brief of appeal against thirst as abusive. Boy, grant it formally and proceed with the hearing!"

"That last bit of bacon . . . pass it over!"

"In the old days, I used to drink my wine to the last drop. Nowatimes I never leave anything in the glass!"

"Here is tripe worth staking money on and chitterlings you'd do well to stay with. Where do they come from? From the black-streaked dun-colored oxen of this region. For God's sake, friends, let us dispatch this meat to the last frazzle!"

"Drink or you'll see what I'll—"

"No, no, enough!"

"Drink, I beg you!"

"Sparrows won't eat unless you tap their tails; I'll not drink unless you treat me politely."

" 'Lagona edatara,' say the Basques: bring me a drink, friend. In all my body there's not a hole or warren for thirst to flee to, without this wine ferreting it out!"

"Ha, this draught will whip my thirst up!"

"Ha, this one will drown mine utterly."

"Let us proclaim to the music of flagons and bottles that whoever has lost his thirst need not look for it here. We have voided it out of doors, thanks to prolonged and laxative libations."

"God in His greatness made the Milky Way, we make the winy; God made the comet, we drink its wine!"

"The Saviour's words are on my lips: '*Sitio,* I thirst!' "

"The stone called asbestos is not more unquenchable than my thirst. Give me to drink: am I not your holy father?"

---

*Capital of Portuguese East Africa when Vasco da Gama made his voyages.

" 'Appetite comes as you eat,' said Bishop Hangest of Le Mans; but thirst vanishes as you drink!"

"Who knows a remedy for thirst?"

"I do! You know how to treat a dog-bite? Well, do exactly the opposite. Always run after a dog and you'll never be bitten; always drink before you're thirsty and you'll never be so!"

"I've caught you: you're asleep. Wake up, man! O steward eternal, keep us from stewing in the juice of sleep. Argus had a hundred eyes to see with; a steward needs a hundred hands, like Briareus, to pour wine indefatigably."

"Let us soak up; to dry thereafter is proud sport!"

"White wine for me, white! Pour up, pour it all out in the devil's name! Pour over here, to the brim: my tongue is peeling!"

"Trink oop, mein frient! Your goot healt', soltchier!"

"Here's to you, mate! Joy and ardor!"

"*O lacryma Christi*, O tears of Christ."

"That wine comes from La Devinière; it's made of the dark grape."

"Oh, what a toothsome little white wine it is!"

"Ay, by my soul it's soft as velvet on the stomach!"

"Ha! that velvet has a rich nap, a smooth texture and a fine finish!"

"Luck to you, comrade!"

"In this game I've done plenty of raising—"

"Your elbow, eh—"

"But I'm several tricks ahead and we'll not lose the hand!"

"*Ex hoc in hoc*: out of my bottle into the mouth. Gentlemen, there is no magic to it: you all saw me! At this job, I'm a past master."

"Ahem, ahem: a massed pastor!"

"Hail to all tosspots! Pity the thirsty!"

"Here, my good lad, fill up my glass. To the top, boy, crown the wine, I pray you!"

"Pour, friend: we'll have my glass red as a cardinal's cape!"

"*Natura abhorret vacuum*—Nature abhors emptiness!"

"Mine is empty: a fly couldn't find a drink in it!"

"Let's drink up like fiddlers, like lords, like Bretons!"

"Polish off this liquor! Come, clean as a whistle!"

"Drink deep, swallow full! Here is a tonic, a sovereign remedy, here is ambrosia!"

—Translated by Jacques Le Clercq

❦

PO CHÜ-I

(CHINESE, 772–846)

Better Come Drink Wine with Me

1

Don't go hide in the deep mountains—
you'll only come to hate it.
Your teeth will ache with the chill of dawn water,
your face smart from the bite of the night frost.
Go off fishing and winds will blow up from the cove;
return from gathering firewood to find snow all over the cliffs.
Better come drink wine with me;
face to face we'll get mellowly, mellowly drunk.

2

Don't go off and be a farmer—
you'll only make yourself miserable.
Come spring and you'll be plowing the lean soil,
twilight and it's time to feed the skinny ox.
Again and again you'll be hit for government taxes,
but seldom will you meet up with a year of good crops.
Better come drink wine with me;
together we'll get quietly, quietly drunk.

3

Don't go climbing up to the blue clouds—*
the blue clouds are rife with passion and hate,
everyone a wise man, bragging of knowledge and vision,
flattening each other in the scramble for merit and power.
Fish get chowdered because they swallow the bait;
moths burn up when they bumble into the lamp.
Better come drink wine with me;
let yourself go, get roaring, roaring drunk.

4

Don't go into the realm of red dust—
it wears out a person's spirit and strength.
You war with each other like the two horns of a snail,
end up with one ox-hair worth of gain.
Put out the fire that burns in your rage,
stop whetting the knife that hides in a smile.
Better come drink wine with me;
we'll lie down peacefully, merrily, merrily drunk.

—Translated by Burton Watson

TUAN TS'AI-HUA

(C H I N E S E , 1 9 3 3 –)

The Feast of "Flower-Pattern" Wine

With his reed raincape hung from his staff, the wandering monk
raised his hand to shield his eyes as he gazed into the distance. On

---

*According to the translator, verses 3 and 4 speak allegorically about the worlds of government (clouds) and the marketplace (dust), eternally opposed and at war.

both sides of the creek, all was a delicate pink. Several steeds with silver bells ambled under willow boughs. The monk waddled ahead toward the densely packed group till he heard the chatter of the girls with their gift-baskets. They were discussing how many sprays of peach blossom they would gather and bring back. The monk hesitated whether to follow the group or turn back to the main road and stood knitting his eyebrows for a while on the Chu-chih Bridge.* A crippled beggar walked up to the bridge, all in tatters and dragging a bamboo cane. Beggars squeezed themselves against the rails on both sides of the bridge. The crippled beggar stopped to ask:

"Isn't it time yet?"

Some of the beggars were dozing, paying no attention to pass-ers-by. The monk, walking toward the middle of the bridge, cast a glance at a woman in a red gown. A beggar guffawed. Peach blossoms filled the stream, and over them twigs and branches of peach intertwined. With his staff held in front of his chest, he was about to break into a rage when a whiff of fragrance came floating in the wind, and he sniffed at it a few times.

"Good smell!" The monk commented with admiration.

The guffawing beggar began to dribble. The crippled beggar stretched his hand to claw at the air a few times as if he wanted to grasp the fragrance with his hands.

"The wine-jars are being opened," he said, "we'll go there as soon as the guests are seated."

The wandering monk suppressed the unhappiness in his heart and walked into the peach wood. On the bank of the creek, there was a horse tied to a red tree while a few others stood around with tethers lying on the ground. A white horse swished its tail. Many green sedan chairs were deposited on the ground in rows. At intervals of every few trees, tables and chairs were set, spread-ing over a square mile in extent, and ladies and gentlemen, all guests at the banquet, were sitting around. The young married women stuck flowers into their pearl brooches and the unmarried girls hid the corners of their mouths behind silk handkerchieves. The male guests had napkins spread over their laps. Cooks and

---

*A bridge in the mountains of Szechuan province.

waiters with dishes or trays in their hands shuttled around among the trees.

"I'd better stay out of sight," the monk said to himself, but he did not move.

"Just this once," he said, spurred on by his appetite. "Except for my initiation scars and shaven head, I am in no way different from those people."

The wine-jars with their carved and painted dragons and phoe-nixes were opened, and the smell of wine filled the air. The monk walked toward the tables spread for the feast; he laid his staff against a small peach tree and leaned on the trunk, his reed raincape swinging slightly.

"Kind sir," he accosted one of the attendants, "is this the Peach Blossom River?"

The man was dipping wine from the jar and funnelling it into bottles to be distributed among the tables. Many others were performing a similar task throughout the area. "No," he answered. "This is only Peach Blossom Creek. We are celebrating the wedding of the landlord's daughter. Unconventional as it is, he chose March for the wedding to entertain his guests in this garden of some three-thousand-and-six-hundred peach trees."

"Oh," said the monk, "I must have taken the wrong road. Otherwise, I'd have reached my destination long ago."

"Where do you come from?" The man stopped dipping and funnelling wine and asked, "Which mountain and which temple are you going to?"

The monk turned to look at the road from which he had just turned off and saw that the road was already buried in drooping flowery branches. "I'm from a broken-down temple on one of the peaks of Omei Mountain," he said. "My master sent me to Puto.* I should have arrived there in February, but the twittering of the swallows has put me off schedule."

"Does a monk from a well-known mountain drink wine?" the man asked. "It seems to me you have come for food and drink."

The monk did not answer; a white-haired old man in a long flower-patterned gown and jacket walked over, his steps light and

---

*Omei and Puto are mountains in Szechuan province famous for Buddhist temples visited by pilgrims. The distance between them is about 150 miles—of pure mountain climbing.

poised and his face glowing with spirits. "On this day of great bliss," the old man said, "we have prepared no vegetarian dishes. If you do not mind eating smelly fish and fat pork, please come and sit at the head of the table."

The attendant stood on one side with drooping shoulders, and the monk could tell that this man was the owner of the place. "The peach trees bloom only once a year!" said the monk.

"Good wine one does not come across many times in a life-time!" the master of the Peach Blossom Creek responded. "Open up a jar specially for his reverence."

The guests stopped laughing and talking, and the monk could feel the eyes of everyone upon him. Becoming red in the face, he made polite gestures trying not to seem presumptuous and then seated himself on a seat alongside the head table. Across the table sat a debonair young man. This young man wore a gown of flowery embroidery with white sleeves turned back at the wrist; holding a twig of peach blossom in one hand, as part of a drinking-game. When the sound of a drum suddenly stopped, the distracted young man had failed to pass the twig to the next guest. The other guests were shouting that he had forfeited and after the monk had settled himself in his seat, the young man was compelled to drink three cups of wine as his punishment.

The young man stood up, graceful as a willow tree in the wind, and raised high his wine cup, saying, "I shall drink every cup to the bottom." He then turned to look at the monk. "But the late-comer should also drink three cups to keep me company."

Under the urging of the master and the guests, the monk consented. He lifted the cup to drink as he stood opposite the young man. He felt his mouth filled with the aged purity and fragrance of the wine.

"Ever since I came down from the mountain, I have come across wine shops in cities and wine stalls in remote villages. I have seen gushing fountains of wine and wine-wagons filling the streets. But, I have passed by them all without even turning my head," said the monk. "What makes me vulnerable to the good smell of this wine?"

The master poured another cup of wine for him, and all the guests at the table stared at him.

"This wine makes one feel that all those nights spent under a lonely green light and inside the ancient walls have been spent in vain; it makes the gods feel the urge to descend into the world of men. This cannot be an ordinary wine. What is its name?"

The drum started again; the drummer was hidden behind a drapery and his sticks fell on the drum rhythmically. The spray of peach blossom was passed to a guest wearing a small green cap, then to a silk merchant, then to a young married woman who drank tea instead of wine, and then to a young girl who wore a white rose in her hair. She passed the spray to the master of Peach Blossom Creek, and the master received the spray with his left hand and passed it on with his right hand. With a smile he said: "Flower-Pattern."

When the spray of peach blossom came into the monk's hand he looked at it and passed the spray to his right hand. Two petals dropped from the spray, one falling beside the cup and the other right into the overflowing cup. "What?" The spray of peach blossom was passed on and the drumbeats grew more and more urgent.

"Flower-Pattern," the host repeated. "That is the name of this wine."

"Ah," the monk said, "I seem to have heard of it."

The debonair young man lifted his chopsticks to pick up a slice of seafood. Placing it in the saucer in front of him, he mixed some sauces with it. "You have heard people talk about it?"

"Hua Tiao,* Mao T'ai, and Ta Ch'ü are well-known wines in the Nine Regions†," said the monk, "and readily available on the market."

The debonair young man broke out into laughter and began to finger the cup in front of him in a playful manner, and all the other guests at the table laughed out loud. The monk asked: "Did I say something wrong?"

"No," said the master of the Peach Blossom Creek, "All kinds of wine are available on the market. Otherwise the drunkard would become even more down on his luck."

As the blossoming spray was passed along, the drum tattoo sounded like a shower. The debonair young man sniffed at it and then passed it on to the guest sitting next to him. Then, he said out loud:

*The Chinese pronunciation of Flower-Pattern Wine.
†Traditional name for Mainland China.

"But Flower-Pattern Wine is unlike such wines as Mao T'ai and Ta Ch'ü, which can be bought at the places where these wines are made. But the Flower-Pattern wine produced in this place is so rare that you can hardly get the genuine thing, no matter how much you are prepared to pay or how far you travel to find it."

"You are exaggerating," said the monk in astonishment.

The young man laughed again, and swallowed his morsel of seafood; he was flushed and excited because of the drum or something else. "It is true," he said, "the wine is made only for weddings. When a woman gives birth to a daughter and the news is reported to her family, her mother provides the rice and her brother invites master-vintners to make wine from the rice. The wine is then sealed into wine-jars with wax or red clay. Famous craftsmen are hired to carve flower patterns on the wine-jars. That is why the wine is called Flower-Pattern."

"Where is the wine stored until the daughter grows up?" asked the monk again.

"After the wine is sealed into the jar, it is transported by horse and wagon to the house of the new-born niece," said the debonair youth. "Trenches of a depth of six to twelve feet are dug beneath the eaves of the house where the niece was born, and the jars are buried in rows. Years of drought and years of war may pass, but not until the time when the girl reaches marriageable age is it allowed to dig up the jars and unseal them."

"Then they remain buried for eighteen to nineteen years?"

"Sometimes even longer." The debonair young man turned to the master of Peach Blossom Creek. "May I speak a word of ill omen?"

"On the day the Flower-Pattern Wine jars are opened, all works are propitious," said the master, clicking his cup with that of the silk merchant.

"In case the girl dies in childhood, the wine is passed on to the next daughter, and the latter will still have her own wine too. So, when this girl is married, she will have two vintages, one belonging to her dead sister and the other belonging to herself."

"Oh," the monk said, "And if she should also die in childhood?"

"It is inherited by a still younger sister. There was one case where the bride had nine vintages of wine."

"If there is no sister left to inherit, what happens?"

"In such cases," said the debonair young man, "there is only a sad adage: The girl has too short a span of life, and people think of the good Flower-Pattern Wine in vain."

When the red flower spray went on its eighth round and fell into the hands of the monk, the drum suddenly stopped. The debonair young man filled up his own cup and also the cup of the guest wearing the small green cap. He pushed both cups to the monk. All the guests in the vicinity turned to look, as the monk first took up his own cup.

"I have not yet extended my congratulations to the master," he said. "Allow me to drink this cup as a toast to you."

The master of Peach Blossom Creek did the same. And the monk said again: "How old is your gracious daughter?"

"Eighteen," the master of the Peach Blossom Creek said. "She looks like a grown-up but she is still no more than a child."

"In that case, this wine has been buried underground for eighteen years?" the monk asked.

"Longer than that," said the debonair young man. "This is the first batch of two hundred jars, and they have been buried for thirty-four years."

"Did she have an older sister then?" the monk asked.

"Yes," the young man said. "When she was sixteen she fell ill with a strange sickness. The most famous physicians from all over China were called in but nobody dared to prescribe. One doctor was still on his way when she died by a small window. Now she lies buried upstream in the midst of the shades of the green willows."

"What a pity," said the woman who drank tea instead of wine. "Such a clever and intelligent lady, and her needle-work was the best for miles around."

"Her death has made this feast possible," said the master, without the slightest touch of sadness on his face. "Please drink freely!" When the monk lifted the cup to his lips, he said again: "Just a while ago one petal fell from the spray and dropped into my cup. . . ."

"It does not matter," the monk went on gaily. "A wine aged for thirty-four years should be sipped with flowers in it."

He had scarcely emptied his cup when there was noise from

somewhere. The few horses tethered under the peach trees bounded lightly, and the white horse began to neigh against the sky. As the guests turned to see, they saw the group of beggars approaching the feast. They were led by the crippled beggar, and they yelled at the top of their lungs their congratulations to the master. The master of Peach Blossom Creek pushed away his chair and walked among the beggars. The guests were all feeling upset and some covered their noses with handkerchieves. The master of Peach Blossom Creek, even in the midst of the shouts and commotion of the beggars, still kept the smile on his face. He promised the beggars that, after the feast, several jars of wine would be opened for them and before they departed they would each be given a present. The crippled beggar told the other beggars to go away but he himself remained among the tables. After he got a drink from one of the waiters, he then began to sing to the accompaniment of two bamboo castanets:

The lotus flowers fall
Fall the lotus flowers
On the northern bank is the master's house
At his daughter's wedding he opens over one thousand jars of wine
And the guests are so numerous that all the pathways are covered with
*   carriages*
In the month of March
Under the red sun
The phoenix-topped and the flower-patterned sedan chair comes from the
*   southern village*
The daughter to be carried out of the house surpasses
The peach blossom in beauty
If you do not drink today, when will you be drinking
This year's blossoms will not bloom next year
Ah, ya, ya, anyone who isn't mad or drunk is a fool
The lotus flowers fall
Fall the lotus flowers
Ah, ya, ya, anyone who isn't mad or drunk is just a fool.

After he stopped singing he swallowed down a huge dipperful of wine and staggered awkwardly away. The monk, enchanted by the song felt hazy and bewildered. The bottom of his cup turned upward slowly; as the cup came in contact with his mouth, the wine with the peach petal in it poured down his throat. He felt a burning sensation in his chest.

He emptied two more cups, and his behavior grew strange. The white horse under the tree was kicking its heels for no reason. With one chopstick and one sprig of peach blossom, he picked up a fat slice of pork and put it into his mouth and chewed at it slowly while looking around with his tipsy eyes at the talking and laughing guests. When the drum sounded again, he passed the sprig to the next guest. All the tables were sounded again, he passed the sprig to the next guest. All the tables were playing the same game, so tens of sprigs were circulating among the guests. The white horse shook himself and his buttocks quivered momentarily. The monk felt ecstatic and lost control of his feelings. He drank another cup, forgetting that he still had a long way to go. He had lived up in the mountains for decades but had never before become so forgetful.

The master of Peach Blossom Creek returned to his own seat, and the girl with a white rose in her hair toasted him. As he filled up the cup, from afar came the sound of string instruments, very much like a crystal stream flowing through clumps of flowers. The debonair young man stood up to watch, and all the sprays of peach blossom stopped abruptly. One of the attendants shouted at the top of his lungs: "The sedan chair has arrived." All the guests one by one laid down their chopsticks, and the drum behind the curtain fell silent. The monk looked with slightly fuddled eyes in the direction from which the sound of music drifted, and he saw a flower-patterned sedan chair carried in; its bright, red curtain was reflected in the creek when it passed over the bridge.

The master of Peach Blossom Creek put down his cup and walked over toward the sedan chair; the woman who drank tea instead of wine and the girl with the white rose stuck into her hair followed the master. And the flowery sedan chair under their guidance passed through the peach orchard and stationed itself to the north of the place of feasting. Some people ran to the house with the news and then the same news was carried back:

"Miss Ch'un-hsin has left the front hall."

"Yin-ti, the servants and the maids of honor, where are they?"

asked the woman who drank tea instead of wine.

"They have gone to conduct the bride here," the messenger said.

A young man wearing a gown of green flowery pattern was led by the master to a seat of honor. He wore a large, red flower printed on his bosom, and even at first sight one could tell that he was the bridegroom. He appeared to be rather restrained in drinking. As he talked and laughed, his face was slightly flushed but he maintained his dashing and bright-tempered spirits. The girl with the white rose in her hair looked at him a few more times and this made the monk's heart beat wildly.

"I do not know who merits greater congratulations," said the guest wearing the small green cap with his eyes on the master of the Peach Blossom Creek. "Miss Ch'un-hsin on marrying such an excellent husband," he turned toward the bridegroom, "or you, on marrying a lady like Miss Ch'un-hsin."

"She is really one in a million," said the woman. "Ever since she was eleven she has seldom come down from her lofty chamber. Aside from inquiring after her parents, she seldom sees anybody."

"I have been with her a few times," the girl with a white rose stuck in her hair said. "I remember two occasions very clearly. Once . . ." Suddenly she suppressed a smile around the corners of her mouth saying: "I had better not mention it."

"It's a glorious day and this is a joyous feast; there's no need to be bashful about differences of station," said the silk merchant. "Why don't you say what you wanted to say?"

The girl with a white rose stuck in her hair looked at the woman, and the woman who drank tea instead of wine said: "Of course. Why don't you speak out?"

"There was one time in March," said the girl with a white rose. "The window of her lofty chamber was thrown open, and cold air drifted in. Sister Ch'un-hsin* was looking at flowers from behind the blinds. A light mist veiled the creek in the early morning. The peach blossoms were in full bloom with their delicate pink. Looked at from the tower Sister Ch'un-hsin was standing in, it appeared to be a dream world. The sun came through the blinds and threw zebra patterns on her. That day she wore a light green dress, with-

* "Sister" here implies friend.

out makeup and with a pale amber pin in her hair. I was dazzled upon seeing her. She asked, 'Why are you staring at me like that?' While she was speaking to me, the wind tucked at the corners of her dress ever so lightly, and her silky hair was uplifted by the wind. And I said: 'Your dress is too thin for the season, isn't it?' And Sister Ch'un-hsin said: 'No.' "

Everybody was lost in thought, imagining that maiden looking at flowers from her tower. The debonair young man had picked up a piece of chicken wing but forgot to put it into his mouth. The guest wearing the little green cap threw his head slightly backward and kept it there. The bridegroom became red in the face. The master of the house and the woman who drank tea instead of wine were smiling. The monk's hand was shaking, agitating the spray of peach blossom between his fingers.

"And the other time?" asked the silk merchant.

The girl again became shy and hesitant, and the woman said: "Go on and finish your story."

"It was August when the wild geese are on the wing," the girl with a white rose started again. "When I think of the time, it seems that I can still hear the cries of the huge geese. Sister Ch'un-hsin was standing for a short while behind the rail on the balcony which overlooked the courtyard. Water dropped from the edge of the roof for it was evening after a rainfall. The setting sun flashed its beams on the evening clouds which appeared in many colors. She was concentrating on the clouds, unaware of my approach. Nothing was in motion except the shadows of the trees which fell across her. That day, she was wearing a white dress, which rippled with delicate colors in the light reflected from the clouds. Her cheeks appeared most tender and radiant and I had never seen such color except that of the lotus flowers in full bloom after rain. When she turned she uttered a quiet 'Ah,' and said: 'When did you come in?' I said, 'I have been standing here for quite a while.' I asked her again: 'What were you thinking?' And Sister Ch'un-hsin replied: 'I have been watching the delicate August clouds, just watching.' As she was saying this, a flight of geese flew over the courtyard, and their cries sounded desolate in the autumn air. Sister Ch'un-hsin knitted her eyebrows, and a cloud of sorrow played across her face."

As she reminisced, all the guests listened intently. The unteth-
ered horses, except for the white one who was still swishing his
tail, stood stock still. The music from the string instruments glided
over the water of the creek. Two men walked ahead unrolling a
red carpet. As the carpet unrolled, the bride, accompanied by the
maids of honor, advanced treading upon it. A boy of fifteen or
sixteen, wearing a broad-brimmed hat, walked beside the bride.
The bride was dressed in a red jacket. On her head was a phoenix
crown with pearls and sparkling precious stones, and her face was
hidden behind a red silk veil. Her eight-pleated skirt was also red
in color; when she moved closer you could hear the silky rustling
of her skirts. All the guests were on their feet and gazed at her
intently. She knelt down on the red carpet and kowtowed to the
master of Peach Blossom Creek to bid farewell, and then she bowed
to thank the guests. The debonair young man smiled and nodded,
and the one with the little green cap bent his head slightly in
acknowledgement. The woman who drank tea instead of wine said
something. The monk let go of the peach spray while gazing at
the neck of the bride; he felt the need to plunge under a roaring
waterfall. Someone set off a string of firecrackers. The exploding
of the firecrackers made the horses jump around. One firecracker
flew all the way to where the white horse was and exploded under
his belly; the white horse jumped and soared over the firecracker
man before he had time to dodge.

The guests shouted and screamed, but the white horse did not
stop. He bumped into a few peach trees and sent the flowers drifting down, then jumped over the phoenix-topped sedan chair but
did no damage. He turned back, leaping and galloping among the
tables, frightening and scattering the guests. The horse swerved
suddenly toward the bride, his hoof beats pounding fiercely. The
servants and maids of honor shrieked and pushed against one another. The bride slipped and stumbled. The red veil dropped from
her face.

The face thus revealed was tender and delicate with a tiny red
mouth, straight nose and delicately arched eyebrows. As the white
horse gambolled madly, she opened her lips slightly to utter a
scream. Her eyes emitted a tremulous and shy light, and she covered her face with her sleeves. Her fingers, protruding from the

sleeves, revealed fingernails as scarlet as her lips.

A wine cup flew from somewhere and hit the horse on his spine, then crashed to the ground; the white horse relaxed his legs and galloped into the wilderness on one side of the peach orchard. The guests returned to their seats, a servant girl picked up the red silk veil and placed it once more over the bride's face. The debonair young man turned back to look but the white horse had already plunged deep into the wilderness. He caught only a glimpse of his white flowing mane, his back and silvery tail, and the sound of his running hooves.

"Ah!" said the guest with the little green cap. "How could that horse turn wild? Has he not been given training?"

"Either the music or the firecrackers or broad vistas of red blossoms must have alarmed him!" said the debonair young man. As he looked back he could see a man still setting off firecrackers under the peach tree.

"Who threw the cup that hit the horse?" asked the master of Peach Blossom Creek. "Otherwise, the horse might have hurt my daughter."

Everybody stared in wonder at each other, but there was no sign of the monk. His seat was empty but the spray of peach blossom was still there. A cup was missing from the place where he had sat. When the guests looked at the peach tree, the staff still leaned on the branches and the green reed cape still hung from the staff. Nothing seemed to have changed.

"Eh?" said the debonair young man. "Why did he leave without bidding goodbye?"

"He came uninvited and left without warning," said the guest in the little green cap. "He is truly a strange fellow."

On the green stone bridge a tumultuous shouting arose. The guests hurried over and found a monk's cloak floating upon the water. Several red peach petals were scattered over the cloak. But the beggars pointed across the creek to a farther place.

As the guests gazed into the distance, they saw a man with bare back and dripping with water. He walked on into the green wilderness without turning his head.

—Translated by Yen Yüan-shu

ⵙ

ŌTOMO TABITO

(J A P A N E S E , 6 5 5 – 7 3 1)

In Praise of Sake,
Thirteen Tanka *

1

Rather than worry
Without result,
One should put down
A cup of rough *sake*.

2

In calling it "sage,"
That splendid sage
Of long ago—how right he was! †

3

What the Seven Sages, too,‡
Long ago craved and craved
Was *sake* above all.

4

Rather than be wise
Churning out words,
Better drink your *sake*,
And weep drunken tears.

---

* A Japanese fixed-verse form composed of alternating lines of five and seven syllables and ending with a seven-syllable line.
† According to the translator, pure sake was secretly called "sage" during the third-century Wei Dynasty prohibition.
‡ Third-century Chin Dynasty sages of the Bamboo Grove.

5

How to speak of it
I know not, yet
The thing I prize
The most is *sake*.

6

Sooner than be a man,
I'd be a *sake* jar,
Soaking in *sake*.

7

O what an ugly sight,
The man who thinks he's wise
And never drinks *sake*!
Give him a good look—
How like an ape he is!

8

Even a priceless jewel—
How can it excel
A cup of rough *sake*?

9

Even jewels that flash
At night—are they like
The draught of *sake*
That frees the mind?

10

Of the ways to play,
In this world of ours,
The one that cheers the heart
Is weeping *sake* tears.

11

If I revel
In this present life,

In the life to come
I may well be a bird,
May well be an insect.

12

"All creatures that live
In the end shall die."
Well, then, while I live
It's pleasure for me.

13

Calm and knowing ways—
These are not for me.
Instead I'd rather weep
Sake-sodden tears!

—*Translated by Geoffrey Bownas*
and Anthony Thwaite

KŌTARŌ TAKAMURA

(JAPANESE, 1883 – 1956)

Plum Wine

The bottle of plum wine made and left by dead Chieko,
dully stagnant with ten years' weight, holds the light,
and in the amber of a wine cup congeals like a jewel-ball.
When alone late at night in the cold time of early spring
please have this, she said.
I think of the one who left this after dying.
Being threatened with the anxiety of a broken mind,
with the distressing idea of ruin before long,
Chieko took care of things around her.
Seven years of madness finished with death.

The fragrant sweetness of this plum wine found in the kitchen
quietly, quietly, I appreciate.
Even the roar of the world of frenzied angry waves
can hardly violate this moment.
When one wretched life is looked straight at
the world just distantly surrounds it.
Now the night wind has stopped.

—*Translated by Edith Marcombe Shiffert and Yōki Sawa*

Y

SHAMS AD-DIN MUHAMMAD [HAFIZ]

(P E R S I A N , 1 3 2 0 ? – 1 3 9 0 ?)

Her hair in disarray, lips laughing;
Drunk in the sweat of revelry
Singing of love, she came, flask in hand.

Disheveled and her clothes rent
Last midnight by my bed she bent;
Her lips curved in regret.

I saw sorrow quarrel in her eyes
As her whispers spoke softly,
"Is our old love asleep?"

Given such a wine before dawn,
A lover is an infidel to love
If he does not drink.

Find no fault, anchorite, with the drinker of dregs,
For on the day of the Covenant
We were given no other gift.

We lift to our lips
Whatever he pours into the wine bowl,
The wine of Paradise or the cup of Hell.

O how many vows of repentance are undone
By the smile of wine and the tresses of a girl
Like the vows of Hafiz?

—Translated by R. M. Rehder

OMAR KHAYYÁM

(PERSIAN, 1048?–1131?)

From *The Rubáiyát*

6

And David's Lips are lock't; but in divine
High piping Pélevi, with "Wine! Wine! Wine!
    Red Wine!"—the Nightingale cries to the Rose
That yellow Cheek of hers to'incarnadine.

11

Here with a Loaf of Bread beneath the Bough,
A Flask of Wine, a Book of Verse—and Thou
    Beside me singing in the Wilderness—
And Wilderness is Paradise enow.

47

And if the Wine you drink, the Lip you press,
End in the Nothing all Things end in—Yes—
    Then fancy while Thou art, Thou art but what
Thou shalt be—Nothing—Thou shalt not be less.

—Translated by Edward Fitzgerald

NATHANIEL HAWTHORNE

(AMERICAN, 1804–1864)

From *The Marble Faun*

"Tomaso, bring some Sunshine!" said he.

The readiest method of obeying this order, one might suppose, would have been to fling wide the green window blinds and let the glow of the summer noon into the carefully shaded room. But, at Monte Beni, with provident caution against the wintry days, when there is little sunshine, and the rainy ones, when there is none, it was the hereditary custom to keep their Sunshine stored away in the cellar. Old Tomaso quickly produced some of it in a small, straw-covered flask, out of which he extracted the cork, and inserted a little cotton wool, to absorb the olive oil that kept the precious liquid from the air.

"This is a wine," observed the Count, "the secret of making which has been kept in our family for centuries upon centuries; nor would it avail any man to steal the secret, unless he could also steal the vineyard, in which alone the Monte Beni grape can be produced. There is little else left me, save that patch of vines. Taste some of their juice, and tell me whether it is worthy to be called Sunshine, for that is its name."

"A glorious name, too!" cried the sculptor.

"Taste it," said Donatello, filling his friend's glass, and pouring likewise a little into his own. "But first smell its fragrance; for the wine is very lavish of it, and will scatter it all abroad."

"Ah, how exquisite!" said Kenyon. "No other wine has a bouquet like this. The flavor must be rare, indeed, if it fulfill the promise of this fragrance, which is like the airy sweetness of youthful hopes, that no realities will ever satisfy!"

This invaluable liquor was of a pale golden hue, like other of the rarest Italian wines, and, if carelessly and irreligiously quaffed, might have been mistaken for a very fine sort of champagne. It was not, however, an effervescing wine, although its delicate piquancy produced a somewhat similar effect upon the palate. Sipping, the guest longed to sip again; but the wine demanded so deliberate a pause, in order to detect the hidden peculiarities and subtile exquisiteness of its flavor, that to drink it was really more a moral than a physical enjoyment. There was a deliciousness in it that eluded analysis, and—like whatever else is superlatively good— was perhaps better appreciated in the memory than by present consciousness.

One of its most ethereal charms lay in the transitory life of the wine's richest qualities; for, while it required a certain leisure and delay, yet, if you lingered too long upon the draught, it became disenchanted both of its fragrance and its flavor.

CHARLES BAUDELAIRE

(FRENCH, 1821–1867)

The Soul of Wine

One evening the wine's soul sung from a flask:
"Man, I send you, dear disinherited one,
From my prison of glass and crimson mask
A song of brotherhood full of light and sun!

I know how much work, sweat and baking sunshine
Are needed to bring me to maturity,
To give me body as a perfect wine.
I will not be ungrateful as you will see,

For I feel an immense joy as I slide
Down the throat of a man worn out with toil,

And his hot breast is a tomb where I can hide
Which pleases me more than the chilly cellar's soil.

Can you hear Sabbath hymns echoing in the cup
And hope that babbles in my throbbing breast?
With elbows on the table, sleeves rolled up
You will glorify me and find yourself at rest.

I will light the eyes of your endearing wife,
Restore to your son his strength and complexion
And as for that frail athlete of life
I am the oil that lubricates his tension.

A vegetable ambrosia, I will descend,
As a precious grain from the eternal Sower
So that our love seeds poetry without end
Which will reach up to God like a rare flower!"

—Translated by John Digby

Y

LI T'AI-PO

(C H I N E S E , 7 0 1 – 7 6 2)

Drinking Alone in the Moonlight

2

If Heaven did not love wine,
There would be no Wine Star in Heaven.
If Earth did not love wine,
There should be no Wine Springs on Earth.
Why then be ashamed before Heaven to love wine.

I have heard that clear wine is like the Sages;
Again it is said that thick wine is like the Virtuous Worthies.
Wherefore it appears that we have swallowed
    both Sages and Worthies.
Why should we strive to be Gods and Immortals?
Three cups, and one can perfectly understand the Great Tao;*
A gallon, and one is in accord with all nature.
Only those in the midst of it can fully comprehend
    the joys of wine;
I do not proclaim them to the sober.

—*Translated by Amy Lowell*
(American, 1874–1925)

Y

MARGARET ST. CLAIR

(A M E R I C A N ,   1 9 1 1 –)

The Wines of Earth

Joe da Valora grew wine in the Napa valley. The growing of pre-
mium wine is never especially profitable in California, and Joe could
have made considerably more money if he had raised soya beans
or planted his acreage in prunes. The paperwork involved in his
occupation was a nightmare to him; he filled out tax and license
forms for state and federal governments until he had moments of
feeling his soul was made out in triplicate, and he worked hard in
the fields too. His son used to ask him why he didn't go into
something easier. Sometimes he wondered himself.

But lovers of the vine, like all lovers, are stubborn and unrea-
sonable men. As with other lovers, their unreasonableness has its
compensations. Joe da Valora got a good deal of satisfaction from
the knowledge that he made some of the best Zinfandel in Cali-

*The way, identified with ultimate reality.

fornia (the Pinot Noir, his first love, he had had to abandon as not coming to its full excellence in his particular part of the Napa valley). He vintaged the best of his wine carefully, slaved over the vinification to bring out the wine's full freshness and fruitiness.

Joe da Valora lived alone. His wife was dead, and his son had married a girl who didn't like the country. Often they came to see him on Sundays, and they bought him expensive gifts at Christmas time. Still, his evenings were apt to be long. If he sometimes drank a little too much of his own product, and went to bed with the edges of things a little blurred, it did him no harm. Dry red table wine is a wholesome beverage, and he was never any the worse for it in the morning. On the nights when things needed blurring, he was careful not to touch the vintaged Zinfandel. It was too good a wine to waste on things that had to be blurred.

Early in December, when the vintage was over and the new wine was quietly doing the last of its fermentation in the storage containers, he awoke to the steady drumming of rain on his roof. Well. He'd get caught up on his bookkeeping. He hoped the rain wouldn't be too hard. Eight of his acres were on a hillside and after every rain he had to do some re-terracing.

About eleven, when he was adding up a long column of figures, he felt a sort of soundless jarring in the air. He couldn't tell whether it was real, or whether he had imagined it. Probably the latter—his hearing wasn't any too good these days. He shook his head to clear it, and began pouring himself some of the unvintaged Zinfandel.

After lunch the rain stopped and the sky grew bright. He finished his wine and started out for a breath of air. As he left the house he realized that he was just a little, little tipsy. Well, that wasn't such a bad way for a vintner to be. He'd go up to the hillside acres and see how they did.

There had been very little soil washing, he saw, inspecting the hillside. The re-terracing would be at a minimum. In fact, most of the soil removal he was doing himself, on the soles of his boots. He straightened up, feeling pleased. Then, ahead of him on the slope he saw four young people, two men and two girls.

Da Valora felt a twinge of annoyance and alarm. What were they doing here? A vineyard out of leaf isn't attractive, and the hillside was well back from the road. He'd never had any trouble with vandals, only with deer. If these people tramped around on the wet earth, they'd break the terracing down.

As he got within speaking distance of them, one of the girls stepped forward. She had hair of an extraordinary copper-gold, and vivid, intensely turquoise, eyes (The other girl had black hair, and the two men were dark blondes.) Something about the group puzzled da Valora, and then he located it. They were all dressed exactly alike.

"Hello," the girl said.

"Hello," da Valora answered. Now that he was near to them, his anxiety about the vines had left him. It was as if their mere proximity—and he was to experience this effect during all the hours they spent with him—both stimulated and soothed his intellect, so that cares and pettinesses dropped away from him, and he moved in a larger air. He seemed to apprehend whatever they said directly, in a deeper way than words are usually apprehended, and with a wonderful naturalness.

"Hello," the girl repeated. "We've come from . . ." somehow the word escaped Joe's hearing, "to see the vines."

"Well, now," said Joe, pleased, "have you seen enough of them? This planting is Zinfandel. If you have, we might go through the winery. And then we might sample a little wine."

Yes they would like to. They would all like that.

They moved beside him in a group, walking lightly and not picking up any of the wet earth on their feet. As they walked along they told him about themselves. They were winegrowers themselves, the four of them, though they seemed so young, in a sort of loose partnership, and they were making a winegrowers' tour of . . . of. . . .

Again Joe's hearing failed him. But he had the fancy that there would never be any conflict of will among the four of them. Their tastes and wishes would blend like four harmonious voices, the women's high and clear, the men's richer and more deep. Yet it seemed to him that the copper-haired girl was regarded with a certain deference by her companions, and he thought, wisely, that he knew the reason. It was what he had so often told his wife— that when a lady really liked wine, when she really had a palate for it, nobody could beat her judgment. So the others respected her.

He showed them through the winery without shame, without pride. If there were bigger wineries than his in the Napa valley, there were smaller ones, too. And he knew he made good wine.

Back in the house he got out a bottle of his vintaged Zinfandel, the best Zindandel he had ever made, for them. It wasn't only that they were fellow growers, he also wanted to please them. It was the '51.

As he poured the dark, fragrant stuff into their glasses he said, "What did you say the name of your firm was? Where did you say you were from?"

"It isn't exactly a firm," the dark-haired girl said, laughing. "And you wouldn't know the name of our home star."

Star? Star? Joe da Valora's hand shook so that he dribbled wine outside the glass. But what else had he expected? Hadn't he known from the moment he had seen them standing on the hillside? Of course they were from another star.

"And you're making a tour?" he asked, putting down the bottle carefully.

"Of the nearer galaxy. We have only a few hours to devote to earth."

They drank. Joe da Valora wasn't surprised when only one of the men, the darker blonde, praised the wine with much vigor. No doubt they'd tasted better. He wasn't hurt—they'd never want to hurt him—or at least not much hurt.

Yet as he looked at the four of them sitting around his dining table—so young, so wise, so kind—he was fired with a sudden honorable ambition. If they were only going to be here a few hours, then it was up to him—since nobody else could do it—it was up to him to champion the wines of earth.

"Have you been to France?" he asked.

"France?" the dark-haired girl answered. So he knew the answer to the question.

"Wait," he told them, "wait. I'll be back." He went clattering down the cellar stairs.

In the cellar, he hesitated. He had a few bottles of the best Pinot Noir grown in the Napa valley; and that meant nobody could question it, the best Pinot Noir grown in California. But which year should he bring? The '43 was the better balanced, feminine, regal, round, and delicate. The '42 was a greater wine, but its inherent imbalance and its age had made it arrive at the state that winemakers call fragile. One bottle of it would be glorious, the next vapid, passe and flat.

In the end, he settled on the '42. He'd take his chances. Just before he left the cellar, he picked up another bottle and carried it up with him. It was something his son had given him a couple of years ago; he'd been saving it for some great occasion. After all, he was championing the wines of *Earth*.

He opened the '42 anxiously. It was too bad he hadn't known about their coming earlier. The burgundy would have benefited by longer contact with the air. But the first whiff of the wine's great nose reassured him; this bottle was going to be all right.

He got clean glasses, the biggest he had, and poured an inch of the wine into them. He watched wordlessly as they took the wine into their mouths, swished it around on their palates, and chewed it, after the fashion of wine-tasters everywhere. The girl with the copper hair kept swirling her glass and inhaling the wine's perfume. He waited tensely for what she would say.

At last she spoke. "Very sound. Very good."

Joe da Valora felt a pang of disappointment whose intensity astonished him. He looked at the girl searchingly. Her face was sad. But she was honest. "Very sound, very good," was all that she could say.

Well, he still had an arrow left in his quiver. Even if it wasn't a California arrow. His hands were trembling as he drew the cork out of the bottle of Romanee-Conti '47 his son had given him. (Where had Harold got it? The wine, da Valora understood, was rare even in France. But the appellation of origin was in order. Harold must have paid a lot for it.)

More glasses. The magnificent perfume of the wine rose to his nose like a promise. Surely this. . . .

There was a long silence. The girl with the dark hair finished her wine and held out her glass for more. At last the other girl said, "A fine wine. Yes, a fine wine."

For a moment Joe da Valora felt he hated her. Her, and the others. Who were these insolent young strangers, to come to Earth, drink the flower, the cream, the very pearl, of earth's vintages, and dismiss it with so slight a compliment? Joe had been drinking wine all his life. In the hierarchy of fine wines, the Zinfandel he made was a petty princeling; the Pinot '42 was a great lord; but the wine he had just given them to drink was the sovereign, the unquestioned emperor. He didn't think it would be possible to grow a better wine on Earth.

The girl with the copper-gold hair got up from the table. "Come to our ship," she said. "Please. We want you to taste the wine we make."

Still a little angry, Joe went with them. The sun was still well up, but the sky was getting overcast. It would rain before night.

The ship was in a hollow behind the hillside vineyard. It was a big silver sphere, flattened at the bottom, that hovered a few feet above the rows of vines. The copper-haired girl took his hand, touched a stud at her belt, and they rose smoothly through the flattened bottom into a sort of foyer. The others followed them.

The ship's interior made little impression on Joe da Valora. He sat down on a chair of some sort and waited while the copper-haired girl went into a pantry and came back with a bottle.

"Our best wine," she said, holding it out for him to see.

The container itself was smaller and squatter than an earth bottle. From it she poured a wine that was almost brownish. He was impressed by its body even in the glass.

He swirled the wine glass. It seemed to him he smelled violets and hazelnuts, and some other perfume, rich and delicate, whose name he didn't know. He could have been satisfied for a half hour, only inhaling the wine's perfume. At last he sipped at it.

"Oh," he said when he had swished it in his mouth, let it bathe his palate, and slowly trickle down his throat. "Oh."

"We don't make much of it," she said, pouring more into his glass. "The grapes are so hard to grow."

"Thank you," he said gratefully. "Now I see why you said, 'A fine wine.'"

"Yes. We're sorry, dear Earth man."

"Don't be sorry," he said, smiling. He felt no sting of inferiority, no shame for Earth. The distance was too great. You couldn't expect Earth vines to grow the wine of paradise.

They were all drinking now, taking the wine in tiny sips, so he saw how precious it was to them. But first one and then the other of them would fill his glass.

The wine was making him bold. He licked his lips, and said,

"Cuttings? Could you . . . give me cuttings? I'd take them to, to the University. To Davis." Even as he spoke he knew how hopeless the words were.

The darker blonde man shook his head. "They wouldn't grow on Earth."

The bottle was empty. Once or twice one of the four had gone to a machine and touched buttons and punched tapes on it. He knew they must be getting ready to go. He rose to leave.

"Good-bye," he said. "Thank you." He held out his hand to them in leave-taking. But all of them, the men too, kissed him lightly and lovingly on the cheek.

"Good-bye, dear human man," the girl with the copper hair said. "Good-bye, good-bye."

He left the ship. He stood at a distance and watched it lift lightly and effortlessly to the height of the trees. There was a pause, while the ship hovered and he wondered anxiously if something had gone wrong. Then the ship descended a few feet and the copper-haired girl came lightly out of it. She came running toward him, one of the small, squat bottles in her hand. She held it out to him.

"I can't take it—" he said.

"Oh, yes. You must. We want you to have it." She thrust it into his hands.

She ran back to the ship. It rose up again, shimmered, and was gone.

Joe da Valora looked at where the ship had been. The gods had come and gone. Was this how Dionysus had come to the Greeks? Divine, bearing a cargo that was divine? Now that they were gone, he realized how much in love with them he had been.

At last he drew a long sigh. He was where he had always been. His life would go on as it always had. Taxes, licenses, a mountain of paperwork, bad weather, public indifference, the attacks of local optionists—all would be as it had been.

But he had the bottle of wine they had given him. He knew there would never, in all his foreseeable life (he was sixty-five), be an event happy enough to warrant his opening it. They had given him one of their last three bottles.

He was smiling as he went back to the house.

ANACREON

(GREEK, 582?–485 B.C.)

Ode 50, The Happy Effects of Wine

SEE! see! the jolly god appears,
His hand a mighty goblet bears;
With sparkling wine full charg'd it flows,
The sovereign cure of human woes.
   Wine gives a kind release from care,
And courage to subdue the fair;
Instructs the cheerful to advance
Harmonious in the sprightly dance,
Hail! goblet, rich with generous wines!
See! round the verge a vine-branch twines.
See! how the mimic clusters roll,
As ready to refil the bowl.
   Wine keeps its happy patients free
From every painful malady;
Our best physician all the year;
Thus guarded, no disease we fear,
No troublesome disease of mind,
Until another year grows kind,
And loads again the fruitful vine,
And brings again our health—new wine.

—*Translated by William Broome*
(English, 1689–1745)

HENRY CAREY

(E N G L I S H , 1 6 8 7 ? – 1 7 4 3)

A Bacchanalian Rant

IN THE BOMBAST STRAIN

Bacchus must now his pow'r resign,
I am the only god of wine:
It is not fit the wretch should be
In competition set with me,
Who can drink ten times more than he.

Make a new world, ye pow'rs divine!
Stock'd with nothing else but wine:
Let wine be earth, and air, and sea,
And let that wine be—all for me.

Let other mortals vainly wear
A tedious life in anxious care;
Let the ambitious toil and think;
Let states or empires swim or sink;
My sole ambition is to *drink*.

❦

THOMAS CHATTERTON

(ENGLISH, 1752 – 1770)

A Bacchanalian

What is war and all its joys?
Useless mischief, empty noise.

What are arms and trophies won?
Spangles glittering in the sun.
Rosy Bacchus, give me wine,
Happiness is only thine!

What is love without the bowl?
'Tis a languor of the soul.
Crowned with ivy, Venus charms;
Ivy courts me to her arms.
Bacchus, give me love and wine,
Happiness is only thine!

❦

HARLEY MATTHEWS

(AUSTRALIAN, 1889 – 1968)

The Vineyard

When I came home from the last war
The place seemed strange—no sign of drought,
Everything green, and it full summer.

My mother closed the window to shut out
The locust just beginning. More
And more the room grew in around me.
Gone the time when a double-drummer
Could lead my mind off. Weeks before
My father, they told me, was dead.
Back now, and a new duty bound me,
All the longed-for calls found me
Unstirred. My mother moved and set
By me the wine she had decanted.
"The bottle of the Hermitage
Put down the day you left," she said.
Few the wine's years, but it had age.
It tasted cool, gentle, and yet
It carried strength, like grown-up men
Moulded by stern events. Outside
The grape-vines that my father planted
In a flush growth of weeds stood dead.
I knew the boy in me had died,
That I had matured like the wine,
When I came home from the war then.

And when I go home from this one?
A vineyard in the years between
I planted, vine and vine and vine.
The way it tilted towards the sun,
And the bush came and sheltered it
Did gladden every vigneron.*
It was pleasure to prune or plough
Up its rows, grey with winter, green
In the heat; a comfort to savour
An old wine when the lamp was lit
At evening; peace to wonder how
The sun of future years would flavour
The juice just rising in the buds
Of some vines newly grafted. Roughened
With work these hands. It only toughened

*Wine grower.

The mind to take the fires and floods,
Spring frosts that blasted every shoot,
And hail that stripped us bare of fruit.
But, sleep-freed from its grief, awoke
The soil, and stirring, yielded all—
Its love, itself—up at our call.

    People, the world over, said meantime,
That peace could be perpetual;
The nations need no more dispute
With armies, once they made a crime
Of war. So it was signed with stroke
Of pen and pen. And most believed
The old stern ways were done; no more
Would men fight, all would be secure,
And those flush, pleasants days endure.

    But harshly they were undeceived:
For all those beliefs there came this war.
Over this planet it has shed
Its dooms and terrors; to this land,
Hardly prepared for fire or drought,
Has brought new forms of death; here bred
Mad hates which make men prisoner
Now.

    The soil's love, my love are wasted.
It lies unploughed, the vine-rows stand
Unpruned; and vintages that were
To grow will never now be tasted.
And this? This had to come about:
Soldier or prisoner he must be
Whose pay is not in currency.
Although the vines are dying out,
The wine is turned to vinegar,
My land sleeps, sleeps, dreams, waiting for
Her own to come home from this war.

—*Anzac Internment Camp,* N.S.W., 1 August 1942

HEINRICH BÖLL

(GERMAN, 1917–1985)

Drinking in Petőcki

The soldier felt he was getting drunk at last. At the same moment it crossed his mind again, very clearly, that he hadn't a single pfennig in his pocket to pay the bill. His thoughts were as crystal-clear as his perception, he saw everything with the utmost clarity: the fat, shortsighted woman sitting in the shadows behind the bar, intent on her crocheting as she chatted quietly to a man with an unmistakably Magyar mustache—a true operetta face, straight from the puszta, while the woman looked stolid and rather German, somewhat too respectable and sedate for the soldier's image of a Hungarian woman. The language they were chatting in was as unintelligible as it was throaty, as passionate as it was strange and beautiful. The room was filled with a dense green twilight from the many close-planted chestnut trees along the avenue leading to the station: a wonderful dense twilight that reminded him of absinthe and made the room exquisitely intimate and cozy. The man with the fabulous mustache, half perched on a chair, looked relaxed and comfortable as he sprawled across the counter.

The soldier observed all this in great detail, at the same time aware that he would not have been able to walk to the counter without falling down. It'll have to settle a bit, he thought, then with a loud laugh shouted "Hey there!", raised his glass toward the woman, and said in German, *"Bitte schön!"* The woman slowly got up from her chair, put aside her crochet work equally slowly, and, carrying the carafe, came over to him with a smile, while the Hungarian also turned round and eyed the medals on the soldier's chest. The woman waddling toward him was as broad as she was tall, her face was kind, and she looked as if she had heart trouble; clumsy pince-nez, attached to a worn black string, balanced on her nose. Her feet seemed to hurt too; while she filled his glass she took the weight off one foot and leaned with one hand on the table. She said something in her dark-toned Hungarian that was doubtless the equivalent of "*Prost*" or "Your very good health," or perhaps even of some affectionate, motherly remark such as old women commonly bestow on soldiers.

The soldier lit a cigarette and drank deeply from his glass. Gradually the room began to revolve before his eyes; the fat proprietress hung somewhere at an angle in the air, the rusty old counter now stood on end, and the Hungarian, who was drinking sparingly, was cavorting about somewhere up near the ceiling like an acrobatic monkey. The next instant everything tilted the other way, the soldier gave a loud laugh, shouted "*Prost!*," took another drink, then another, and lit a fresh cigarette.

The door opened and in came another Hungarian, fat and short, with a roguish onion face and a few dark hairs on his upper lip. He let out a gusty sigh, tossed his cap onto a table, and hoisted himself onto a chair by the counter. The woman poured him some beer. . . .

The gentle chatter of the three at the counter was wonderful, like a quiet humming at the edge of another world. The soldier took another gulp of wine, put down his empty glass, and everything resumed its proper place.

The soldier felt almost happy as he raised his glass again, repeating with a laugh, "*Bitte schön!*"

The woman refilled his glass.

I've had almost ten glasses of wine, the soldier thought. I'll stop now, I'm so gloriously drunk that I feel almost happy. The green twilight thickened, the farther corners of the bar were already filled with impenetrable deep-blue shadows. What a crime, thought the soldier, that there are no lovers here. It would be a perfect spot for lovers, in this wonderful green-and-blue twilight. What a crime, he thought, as he pictured all those lovers somewhere out there in the world who had to sit around or chase around in the bright light, while here in the bar there was a place where they could talk, drink wine, and kiss. . . .

Christ, thought the soldier, there ought to be music here now, and all these wonderful dark-green and dark-blue corners ought to be full of lovers—and I would sing a song. You bet I'd sing a song. I feel very happy, and I would sing those lovers a song, then I'd really quit thinking about the war; now I'm always thinking a little bit about this damn war. Then I'd quit thinking about it altogether.

He looked closely at his watch: seven-thirty. He still had twenty minutes. He drank long and deep of the dry, cool wine, and it was almost as if someone had given him stronger spectacles: now everything looked closer and clearer and very solid, and he felt himself becoming gloriously, beautifully, almost totally drunk. Now he saw that the two men at the counter were poor, either laborers or shepherds, in threadbare trousers, and that their faces were tired and terribly submissive in spite of the dashing mustache and the wily onion look. . . .

Christ, thought the soldier, how horrible it was back there when I had to leave, so cold, and everything bright and full of snow, and we still had a few minutes left and nowhere was there a corner, a wonderful, dark, human corner where we could have kissed and embraced. Everything had been bright and cold. . . .

"Bitte schön!" he shouted to the woman; then, as she approached, he looked at his watch: he still had ten minutes. When the woman started to fill his half-empty glass, he held his hand over it, shook his head with a smile, and rubbed thumb and forefinger together. "Pay," he said, "how many pengös?"*

He very slowly took off his jacket, slipped off the handsome gray turtleneck sweater, and laid it beside him on the table in front of the watch. The men at the counter had stopped talking and were looking at him, the woman also seemed startled. Very carefully she wrote a 14 on the tabletop. The soldier placed his hand on her fat, warm forearm, held up the sweater with the other, and asked with a laugh, "How much?" Rubbing thumb and forefinger together again, he added, "Pengös."

The woman looked at him and shook her head, but he went on shrugging his shoulders and indicating that he had no money until she hesitantly picked up the sweater, turned it over, and carefully examined it, even sniffed it. She wrinkled her nose a little, then

---

*A Hungarian monetary unit.

smiled and with a pencil quickly wrote a "30" next to the "14."
The soldier let go of her warm arm, nodded, raised his glass, and
took another drink.

As the woman went back to the counter and eagerly began talking to the men in her throaty voice, the soldier simply opened his
mouth and sang. He sang "When the Drum Roll Sounds for Me,"
and suddenly realized he was singing well—singing well for the
first time in his life; at the same time he realized he was drunker
again, that everything was gently swaying. He took another look
at his watch and saw he had three minutes in which to sing and
be happy, and he started another song, "Innsbruck, I Must Leave
You." Then with a smile he took the money the woman had placed
in front of him and put it in his pocket. . . .

It was quite silent now in the bar. The two men with the
threadbare trousers and the tired faces had turned toward him,
and the woman had stopped on her way back to the counter and
was listening quietly and solemnly, like a child.

The soldier finished his wine, lit another cigarette, and knew he
would walk unsteadily. But before he left he put some money on
the counter and, with a *"Bitte schön,"* pointed to the two men. All
three stared after him as he at last opened the door and went out
into the avenue of chestnut trees leading to the station, the avenue
that was full of exquisite dark-green, dark-blue shadows where a
fellow could have put his arms around his girl and kissed her good-
bye. . . .

—Translated by Leila Venneintz

Y

CARL ZUCKMAYER

(GERMAN, 1896–1977)

To the Red Stains on a Tablecloth in a French Restaurant

I look at you with solemn joy wine stain
And push away the plate from where you hide.
My first gulp toasts that unknown man who came
Before me and savored his meal at this table side.

From your lilac edge, spreading like watery hands,
With a drunkard's dreamy tender gaze you peep,
Resembling silhouettes of foreign lands,
Madagascar perhaps or Mozambique.

With golden crumbs my place is now strewn carelessly,
Bread someone slowly broke as he sat down to dine.
O melodic land of fragrant burgundy
Are you the self-same principality
That once a King assessed in tuns of wine?

You, ripe land, played upon by evening light,
Through the mellow tongue of a many-colored prism
Where toll-collectors painted and geniuses took flight,
Where even God forgot his heavenly vision.

I saw you torn with steel and sealed in blood.
In agony I lay on your body, a churned slough.
Perhaps it was he who shot at me from the mud
That friendly, stout sommelier who serves me now.

Ah, didn't I drink at your well of tears that day?
Didn't I see you almost die, suffer your pain?
Sister land! I raise my glass to your rich clay,
At your threshold I kiss every wine and blood stain.

He fills my glass, wine sparkles from the bottle's mouth.
So drink tablecloth, drink like a thirsty dog.
This foreigner bows to your honor, your sunlit south
And heads reluctantly north into the fog.

—Translated by John Digby

GUILLAUME APOLLINAIRE

(F R E N C H , 1 8 8 0 – 1 9 1 8)

Rhenish Night

My glass brimming with wine shimmers like a flame
Listen to the boatman's song its slow sad beat
That tells how seven girls in moonlight came
To twist their long green hair falling to their feet

Arise sing higher while dancing in a ring
Out sing out sing the drifting boatman's song
And bring me those bewitching ladies who sing
Their gazes fixed and twisted braids so long

The drunken Rhine mirrors vineyards in its stream
Where all the golden nights reflect and tremble
A death-rattle song forever like a dream
Of summer spells that green-haired maids dissemble

My glass is broken like a burst of laughter

—Translated by John Digby

Y

EDGAR ALLAN POE

(AMERICAN, 1809–1849)

The Cask of Amontillado

The thousand injuries of Fortunato I had borne as I best could, but when he ventured upon insult I vowed revenge. You, who so well know the nature of my soul, will not suppose, however, that I gave utterance to a threat. *At length* I would be avenged; this was a point definitely settled—but the very definitiveness with which it was resolved precluded the idea of risk. I must not only punish but punish with impunity. A wrong is unredressed when retribution overtakes its redresser. It is equally unredressed when the avenger fails to make himself felt as such to him who has done the wrong.

It must be understood that neither by word nor deed had I given Fortunato cause to doubt my good will. I continued, as was my wont, to smile in his face, and he did not perceive that my smile *now* was at the thought of his immolation.

He had a weak point—this Fortunato—although in other regards he was a man to be respected and even feared. He prided himself on his connoisseurship in wine. Few Italians have the true virtuoso spirit. For the most part their enthusiasm is adopted to suit the time and opportunity, to practice imposture upon the British and Austrian millionaires. In painting and gemmary, Fortunato, like his countrymen, was a quack, but in the matter of old wines he was sincere. In this respect I did not differ from him materially; I was skillful in the Italian vintages myself, and bought largely whenever I could.

It was about dusk, one evening during the supreme madness of the carnival season, that I encountered my friend. He accosted me with excessive warmth, for he had been drinking much. The man wore motley. He had on a tight-fitting parti-striped dress, and his head was surmounted by the conical cap and bells. I was so pleased to see him that I thought I should never have done wringing his hand.

I said to him—"My dear Fortunato, you are luckily met. How remarkably well you are looking to-day. But I have received a pipe* of what passes for Amontillado, and I have my doubts."

"How?" said he. "Amontillado? A pipe? Impossible! And in the middle of the carnival!"

"I have my doubts," I replied; "and I was silly enough to pay the full Amontillado price without consulting you in the matter. You were not to be found, and I was fearful of losing a bargain."

"Amontillado!"

"I have my doubts."

"Amontillado!"

"And I must satisfy them."

"Amontillado!"

"As you are engaged, I am on my way to Luchresi. If any one has a critical turn, it is he. He will tell me—"

"Luchresi cannot tell Amontillado from Sherry."

"And yet some fools will have it that his taste is a match for your own."

"Come, let us go."

"Whither?"

"To your vaults."

"My friend, no; I will not impose upon your good nature. I perceive you have an engagement. Luchresi—"

"I have no engagement;—come."

"My friend, no. It is not the engagement, but the severe cold with which I perceive you are afflicted. The vaults are insufferably damp. They are encrusted with niter."

"Let us go, nevertheless. The cold is merely nothing. Amontillado! You have been imposed upon. And as for Luchresi, he cannot distinguish Sherry from Amontillado."

Thus speaking, Fortunato possessed himself of my arm; and putting on a mask of black silk and drawing a *roquelaire*† closely about my person, I suffered him to hurry me to my palazzo.

* A cask holding 126 gallons.
† An eighteenth-century man's cloak.

There were no attendants at home; they had absconded to make merry in honor of the time. I had told them that I should not return until the morning, and had given them explicit orders not to stir from the house. These orders were sufficient, I well knew, to insure their immediate disappearance, one and all, as soon as my back was turned.

I took from their sconces two flambeaux, and giving one to Fortunato, bowed him through several suites of rooms to the archway that led into the vaults. I passed down a long and winding staircase, requesting him to be cautious as he followed. We came at length to the foot of the descent, and stood together on the damp ground of the catacombs of the Montresors.

The gait of my friend was unsteady, and the bells upon his cap jingled as he strode.

"The pipe?" said he.

"It is farther on," said I; "but observe the white web-work which gleams from these cavern walls."

He turned towards me, and looked into my eyes with two filmy orbs that distilled the rheum of intoxication.

"Niter?" he asked, at length.

"Niter," I replied. "How long have you had that cough?"

"Ugh! ugh! ugh!—ugh! ugh! ugh!—ugh! ugh! ugh! ugh! ugh! ugh!—ugh! ugh! ugh!"

My poor friend found it impossible to reply for many minutes.

"It is nothing," he said, at last.

"Come," I said, with decision, "we will go back; your health is precious. You are rich, respected, admired, beloved; you are happy, as once I was. You are a man to be missed. For me it is no matter. We will go back; you will be ill, and I cannot be responsible. Besides, there is Luchresi—"

"Enough," he said; "the cough is a mere nothing; it will not kill me. I shall not die of a cough."

"True—true," I replied; "and, indeed, I had no intention of alarming you unnecessarily—but you should use all proper caution. A draft of this Medoc will defend us from the damps."

Here I knocked off the neck of a bottle which I drew from a long row of its fellows that lay upon the mold.

"Drink," I said presenting him the wine.

He raised it to his lips with a leer. He paused and nodded to me familiarly, while his bells jingled.

"I drink," he said, "to the buried that repose around us."

"And I to your long life."

He again took my arm, and we proceeded.

"These vaults," he said, "are extensive."

"The Montresors," I replied, "were a great and numerous family."

"I forget your arms."

"A huge human foot d'or,* in a field azure; the foot crushes a serpent rampant whose fangs are imbedded in the heel."

"And the motto?"

"Nemo me impune lacessit?"†

"Good!" he said.

The wine sparkled in his eyes and the bells jingled. My own fancy grew warm with the Medoc. We had passed through long walls of piled skeletons, with casks and puncheons intermingling, into the inmost recesses of the catacombs. I paused again, and this time I made bold to seize Fortunato by an arm above the elbow.

"The niter!" I said; "see, it increases. It hangs like moss upon the vaults. We are below the river's bed. The drops of moisture trickle among the bones. Come, we will go back ere it is too late. Your cough——"

"It is nothing," he said; "let us go on. But first, another draft of the Medoc."

I broke and reached him a flagon of De Grâve. He emptied it at a breath. His eyes flashed with a fierce light. He laughed and threw the bottle upward with a gesticulation I did not understand.

I looked at him in surprise. He repeated the movement—a grotesque one.

"You do not comprehend?" he said.

"Not I," I replied.

"Then you are not of the brotherhood."

"How?"

"You are not of the masons."‡

"Yes, yes," I said; "yes, yes."

*Of gold
†No one provokes me with impunity.
‡Freemasons, here deliberately confused with stonemasons.

"You? Impossible! A mason?"

"A mason," I replied.

"A sign," he said, "a sign."

"It is this," I answered, producing from beneath the folds of my *roquelaire* a trowel.

"You jest," he exclaimed, recoiling a few paces. "But let us proceed to the Amontillado."

"Be it so," I said, replacing the tool beneath the cloak and again offering him my arm. He leaned upon it heavily. We continued our route in search of the Amontillado. We passed through a range of low arches, descended, passed on, and descending again, arrived at a deep crypt, in which the foulness of the air caused our flambeaux rather to glow than flame.

At the most remote end of the crypt there appeared another less spacious. Its walls had been lined with human remains, piled to the vault overhead, in the fashion of the great catacombs of Paris. Three sides of this interior crypt were still ornamented in this manner. From the fourth the bones had been thrown down, and lay promiscuously upon the earth, forming at one point a mound of some size. Within the wall thus exposed by the displacing of the bones, we perceived a still interior crypt or recess, in depth about four feet, in width three, in height six or seven. It seemed to have been constructed for no especial use within itself, but formed merely the interval between two of the colossal supports of the root of the catacombs, and was backed by one of their circumscribing walls of solid granite.

It was in vain that Fortunato, uplifting his dull torch, endeavored to pry into the depth of the recess. Its termination the feeble light did not enable us to see.

"Proceed," I said; "herein is the Amontillado. As for Luchresi—"

"He is an ignoramus," interrupted my friend, as he stepped unsteadily forward, while I followed immediately at his heels. In an instant he had reached the extremity of the niche, and finding his progress arrested by the rock, stood stupidly bewildered. A moment more and I had fettered him to the granite. In its surface were two iron staples, distant from each other about two feet, horizontally. From one of these depended a short chain, from the other a padlock. Throwing the links about his waist, it was but the work of a few seconds to secure it. He was too much astounded to resist. Withdrawing the key I stepped back from the recess.

"Pass your hand," I said, "over the wall; you cannot help feeling the niter. Indeed it is *very* damp. Once more let me *implore* you to return. No? Then I must positively leave you. But I must first render you all the little attentions in my power."

"The Amontillado!" ejaculated my friend, not yet recovered from his astonishment.

"True," I replied; "the Amontillado."

As I said these words I busied myself among the pile of bones of which I have before spoken. Throwing them aside, I soon uncovered a quantity of building stone and mortar. With these materials and with the aid of my trowel, I began vigorously to wall up the entrance of the niche.

I had scarcely laid the first tier of the masonry when I discovered that the intoxication of Fortunato had in a great measure worn off. The earliest indication I had of this was a low moaning cry from the depth of the recess. It was *not* the cry of a drunken man. There was then a long and obstinate silence. I laid the second tier, and the third, and the fourth; and then I heard the furious vibrations of the chain. The noise lasted for several minutes, during which, that I might hearken to it with the more satisfaction, I ceased my labors and sat down upon the bones. When at last the clanking subsided, I resumed the trowel, and finished without interruption the fifth, the sixth, and the seventh tier. The wall was now nearly upon a level with my breast. I again paused, and holding the flambeaux over the mason-work, threw a few feeble rays upon the figure within.

A succession of loud and shrill screams, bursting suddenly from the throat of the chained form, seemed to thrust me violently back. For a brief moment I hesitated, I trembled. Unsheathing my rapier, I began to grope with it about the recess; but the thought of an instant reassured me. I placed my hand upon the solid fabric of the catacombs, and felt satisfied. I reapproached the wall. I replied to the yells of him who clamored. I reechoed, I aided, I surpassed them in volume and in strength. I did this, and the clamorer grew still.

It was now midnight, and my task was drawing to a close. I had completed the eighth, the ninth and the tenth tier. I had finished a portion of the last and the eleventh; there remained but a single stone to be fitted and plastered in. I struggled with its weight; I placed it partially in its destined position. But now there came from out the niche a low laugh that erected the hairs upon my head. It was succeeded by a sad voice, which I had difficulty in recognizing as that of the noble Fortunato. The voice said—

"Ha! ha! ha!—he! he! he!—a very good joke, indeed—an excellent jest. We will have many a rich laugh about it at the palazzo—he! he! he!—over our wine—he! he! he!"

"The Amontillado!" I said.

"He! he! he!—he! he! he!—yes, the Amontillado. But is it not getting late? Will not they be awaiting us at the palazzo, the Lady Fortunato and the rest? Let us be gone."

"Yes," I said, "let us be gone."

"For the love of God, Montresor!"

"Yes," I said, "for the love of God!"

But to these words I hearkened in vain for a reply. I grew impatient. I called aloud—

"Fortunato!"

No answer. I called again—

"Fortunato!"

No answer still. I thrust a torch through the remaining aperture and let it fall within. There came forth in return only a jingling of the bells. My heart grew sick; it was the dampness of the catacombs that made it so. I hastened to make an end of my labor. I forced the last stone into its position; I plastered it up. Against the new masonry I re-erected the old rampart of bones. For the half of a century no mortal has disturbed them. *In pace requiescat!* *

* Rest in peace.

Y

ALEXANDER BROME

(E N G L I S H , 1 6 2 0 – 1 6 6 6)

On Canary

1

Of all the rare juices.
That *Bacchus or Caeres* * produces,
   There's none that I can, nor dare I
   Compare with the princely Canary
      For this is the thing
      That a fancy infuses,
         This first got a King,
         And next the nine Muses,
'Twas this made old Poets so sprightly to sing.
   And fill all the world with the glory and fame on't.
They *Helicon* call'd it and the *Thespian* spring,†
   But this was the drink, though they knew not the name on't.

2

Our Sider and Perry,
May make a man mad but not merry:
   It makes people windmill-pated,
   And with crackers sophisticated,
      And your hopps, yest, and malt.
      When they're mingled together.
         Makes our fancies to halt,
         Or reel any whether.
It stuffs up our brains with froth and with yest,
   That if one would write but a verse for a Belman,

---

* God of wine and goddess of grain.
† Water from the legendary Greek spring that came into being when the ground was struck by the hoof of the winged horse Pegasus.

He must study till Christmas for an eight shilling jest,
   These liquors won't raise, but drown, and o're-whelme man.

3

  Our drousy Matheglin
Was only ordain'd to enveigle in.
   The Novice that knowes not to drink yet,
   But is fudled before he can think it;
     And your Claret and White,
    Have a Gunpowder fury,
     They're of the *French* spright,
    But they wont long endure you.
And your holiday Muscadine, Allegant and Tent,
   Have only this property and vertue that's fit in't;
They'l make a man sleep till a preachment be spent,
   But we neither can warm our blood nor our wit in't.

4

  The Bagrag* and Rhenish
You must with ingredients replenish;
   'Tis a wine to please Ladies and toyes with
   But not for a man to rejoyce with.
     But 'tis Sack makes the sport.
    And who gains but that flavour,
     Though an Abbesse he court,
    In his highshoes he'l have her.
'Tis this that advances the drinker and drawer,
   Though the father came to Town in his hobnails and leather,
He turns it to velvet, and brings up an Heir,
   In the Town in his chain, in the field with his feather.

*An English corruption, but of what wine we have been unable to discover.

♀

WILLIAM SHAKESPEARE

(ENGLISH, 1564 – 1616)

From *Henry IV, Part 2,* IV, iv

FALSTAFF. I would you had [but] the wit; 'twere better than your dukedom. Good faith, this same young sober-blooded boy doth not love me, nor a man cannot make him laugh; but that's no marvel, he drinks no wine. There's never none of these demure boys come to any proof; for thin drink doth so over-cool their blood, and making many fish-meals, that they fall into a kind of male green-sickness;* and then, when they marry, they get wenches. They are generally fools and cowards; which some of us should be too, but for inflammation. A good sherris-sack hath a two-fold operation in it. It ascends me into the brain; dries me there all the foolish and dull and crudy vapours which environ it; makes it apprehensive, quick, forgetive, full of nimble, fiery, and delectable shapes; which, delivered o'er to the voice, the tongue, which is the birth, becomes excellent wit. The second property of your excellent sherris is, the warming of the blood; which, before cold and settled, left the liver white and pale, which is the badge of pusillanimity and cowardice; but the sherris warms it and makes it course from the inwards to the parts extremes. It illumineth the face, which as a beacon gives warning to all the rest of this little kingdom, man, to arm; and then the vital commoners and inland petty spirits muster me all to their captain, the heart, who, great and puff'd up with this retinue, doth any deed of courage; and this valour comes of sherris. So that skill in the weapon is nothing without sack, for that sets it a-work; and learning a mere hoard of gold kept by a devil, till sack commences it and set it in act and use.

* An anemia associated with young girls going through puberty.

Hereof comes it that Prince Harry is valiant; for the cold blood he did naturally inherit of his father, he hath, like lean, sterile, and bare land, manured, husbanded, and till'd with excellent endeavour of drinking good and good store of fertile sherris, that he is become very hot and valiant. If I had a thousand sons, the first humane principle I would teach them should be, to forswear thin potations and to addict themselves to sack.

♆

GAIUS VALERIUS CATULLUS

(ROMAN, 84?–54? B.C.)

27

Listen kid, go bring us something
decent to drink, you heard the lady,
she's smashed as a grape and wants
good old Falernian bubbly, the best.
Get out, water, you kill the wine,
move, go chase the squares instead,
over here we do only serious drinking.

—Translated by Carl Sesar

GEORGE GORDON, LORD BYRON

(ENGLISH, 1788–1824)

From *Don Juan*

178

And the small ripple spilt upon the beach
　　Scarcely o'erpassed the cream of your champagne,
When o'er the brim the sparkling bumpers reach,
　　That spring-dew of the spirit! the heart's rain!
Few things surpass old wine; and they may preach
　　Who please,—the more because they preach in vain,—
Let us have Wine and Woman, Mirth and Laughter,
Sermons and soda-water the day after.

179

Man, being reasonable, must get drunk;
　　The best of Life is but intoxication:
Glory, the Grape, Love, Gold, in these are sunk
　　The hopes of all men, and of every nation;
Without their sap, how branchless were the trunk
　　Of Life's strange tree, so fruitful on occasion!
But to return,—Get very drunk, and when
You wake with headache—you shall see what then!

180

Ring for your valet—bid him quickly bring
　　Some hock and soda-water, then you'll know
A pleasure worthy Xerxes the great king;*
　　For not the blest sherbet, sublimed with snow,

*Persian king (486–465 B.C.); the son of Darius I, he was known for his invasion of
Greece across the Hellespont, which Byron also swam.

Nor the first sparkle of the desert-spring,
  Nor Burgundy in all its sunset glow,
After long travel, Ennui, Love, or Slaughter,
Vie with that draught of hock and soda-water!

♥

JOHN DIGBY

(E N G L I S H , 1 9 3 8 –)

The Champagne Party

Mr. Norman Bernstein coughed two or three times in order to gain everybody's attention in the room. The buzzing of the conversation gradually died away like flies departing at the end of summer, and silence fell like a heavy curtain. Although the room was not large over a hundered people crowded into it, smoking and talking all at once. Now suddenly there was silence and all eyes turned to Mr. Norman Bernstein, the managing director of the Botkin Tie and Shirt Manufacturers. Among the group of younger men and women stood Chuck Goodnough. He was in his late seventies. He was standing in the middle of the room sheepishly, with his arms behind his back.

Mr. Norman Berstein walked to the front of his antique oak desk and stood in front of it. He was clutching a couple of sheets of paper in his hand. He looked at Chuck Goodnough and smiled.

"I spent the last week writing this farewell speech about Chuck," he said, "but, hell, everyone knows what a great guy he is. Anyway, I don't think he wants to hear about himself, and to tell you the truth, the speech wasn't so hot."

A few people laughed.

"Chuck," he said, addressing everyone in the room, "started with my grandfather, the old, terrible tyrant Mr. Botkin over sixty-three years ago and has remained with the firm through its ups and downs for all those years."

There was loud and long applause. Chuck smiled and looked down at his shoes, a little embarrassed.

"What can I say?" Mr. Bernstein threw his hands up into the air. "You've been a good, no a great porter and a faithful employee for all those years. First Chuck," Mr. Bernstein reached into his inside pocket and handed him a plain envelope, "here's something to keep the wolves away from the door."

Chuck held the envelope, unable to make up his mind whether or not to open it.

"Open it if you want," said Mr. Bernstein.

Chuck dutifully opened the envelope. It was a cheque for four figures and a pretty good one at that. Chuck's mind toyed with what he was going to do with all that money.

"Now," Mr. Bernstein continued, "the staff have their own surprise for you as well."

Mr. Bernstein stretched up on tiptoes and looked at the back of the room. "Wheel it in boys," he said loudly.

The office door swung open and three men pushed a shape under a white sheet into the middle of the room where Chuck stood. Leslie, one of the younger secretaries, came and stood in front of the shape.

"Chuck," she said, trying hard to control her tears, "this is from everyone because we love you." Many people started clapping.

"Come on, Chuck, pull it off and take a peek," said Mr. Grayson, the senior accountant.

He moved closer to the white shape, gingerly lifted a corner, and then pulled off the white sheet. Under it was a very expensive leather recliner. Even Chuck was a little surprised.

"Something to put your feet up on," offered Mr. Bernstein. "Try it for size."

The people in the room fell silent again. Chuck walked around to the front of the recliner and as soon as he settled into it the whole crowd burst into song, "For he's a jolly good fellow."

His cheque and envelope fell to the floor. No one noticed it. Someone shouted out, "Speech!"

"No," said Mr. Bernstein firmly, waving one hand in the air. "One more surprise—a champagne party for Chuck."

Leslie noticed the cheque and envelope on the floor. She picked them up and placed them in Chuck's coat pocket as he sat in the recliner smiling and obviously pleased with all the fuss that people were making over him. Suddenly around the room soft explosions were occurring as people were popping champagne bottles. Then they gathered around Chuck and raised their glasses. Chuck held his in his hand. Mr. Bernstein toasted him.

"To Chuck. Thanks for sixty-three years." Everyone waited until Chuck took the first sip. He raised his glass to his mouth and before he swallowed the room echoed with the words, "to Chuck."

As he lolled like an emperor in his recliner in the middle of the room everyone pressed around him and tried to talk to him at the same time. They were firing questions at him about the old days. There were so many questions that Chuck couldn't start telling them apart, let alone giving any answers.

"Ever had champagne before, Chuck?" asked Mr. Bernstein.

"No, I don't think I have," answered Chuck.

"It tickles your nose," said Mr. Bernstein smiling.

"That it does," replied Chuck.

"Ever had champagne before?" The words echoed in his head. Chuck had lied out of politeness or to hide a memory from well over thirty years ago.

Leslie, the secretary, kept filling his glass with champagne. Chuck had always liked her because she reminded him of his own Emily, his wife who had died over twenty-five years ago.

"How long have you been with the firm, Leslie?" asked Chuck, sipping his champagne.

"Ten years."

"I remember the week you started."

"I was only going to stay a few months, I thought of it as a temporary job."

"How's your young man?" Chuck asked suddenly.

"He's doing well—senior electrician now uptown, and next year is the big one. Next year we get married."

She filled his glass again. "The bubbles tickle my nose," said Chuck. They both laughed.

"Leslie," said Chuck.

"Yes," answered the girl, bending over him.

"Slip around and see me before you get married, won't you," asked Chuck.

"Slip around and see you? You're coming to the wedding. Chuck Goodnough, chief guest of honor."

"You don't want an old fogey like me around all those young folks."

"O yes I do, you wait and see," Leslie said. "You're coming to the wedding and that's final," she said smiling.

The people in the room were gradually breaking into their own small intimate groups and carrying on their own conversations with low laughter. Every now and then someone came up and filled Chuck's glass, and he kept drinking from it. As he stretched back in his new recliner the voices in the room were getting softer and slowly fading. He tried hard to keep his eyes open, but what with the champagne and the heat he was drifting into sleep.

As he drifted into sleep for some unknown reason he awakened the memory of his twenty-fifth wedding anniversary. He had purchased his wife a pair of gold earrings, but he considered that the gift was not enough for so special an occasion . . . but what? He must have thought for several days and at last he found the answer—a champagne dinner at a swanky restaurant uptown. The problem was that it would cost money and plenty of it, and where was he going to get hold of such an amount? He thought he could walk to and from work and save a little that way, he could skip lunch, but it still wasn't enough. Not enough really to save in time enough to take Emily out on their anniversary.

He struggled, it was a terrible idea; the petty cash was never locked in the drawer at night. He had only to linger late at work and . . . He would borrow it just for a few weeks and pay it back. He certainly could save the amount he was contemplating borrowing. It was an appealing idea and the firm wouldn't miss it for a few weeks. It was indeed tempting; he took the risk.

That evening at the restaurant when the waiter came over to their table and asked if they wanted a cocktail Chuck said, "No thank you. We'll have champagne, and the best."

"Champagne," said Emily, "a little expensive."

"Nothing's too expensive for my wife," said Chuck, reaching over the table, taking her hand, and kissing it quickly.

That evening they drank a magnum of champagne and danced to the slow music.

"It's only once you have a twenty-fifth anniversary," said Chuck.

He had promised himself that he was going to return the money that very month. He considered it nothing more than a loan.

When he had saved the amount that he had taken from the petty cash he lingered late at the office, but to his surprise the box was locked. It had never been locked before and it was never left unlocked again. The following week one of the secretaries suddenly departed. She left mid-week. At first Chuck never realized that she might have been dismissed because of the missing money. For years the thought haunted him. He couldn't help thinking that the poor girl might have been called in to old Mr. Botkin's office and been dismissed or actually accused of taking the money.

Chuck was determined to tell old Mr. Botkin the truth, but every time he was about to march into the office his nerve failed him. He knew that he would probably get dismissed, and he just couldn't face Emily; he felt too ashamed and the pleasant memory of the night of their anniversary would have turned sour.

It's not too late now, Chuck thought, as he struggled out of sleep. Yes, he would tell Mr. Bernstein, old Mr. Botkin's grandson and confess it once and for all. Chuck was sure that Mr. Bernstein would understand. He was kindly and he had a soft spot for him. He could take the cash out of the cheque—with interest of course. And anyway he never really needed all that money, too much for an old man. He wanted to give Leslie a big chunk of it as a wedding present. He had no need of it and no one to leave it to. Yes Mr. Bernstein, said Chuck to himself, I want a few words with you.

As Chuck was about to struggle up from the recliner Emily waltzed into the room, yes Emily, fresh and as beautiful as the day that they married.

"O Chuck," Emily said, as she bent over him and kissed his forehead, "It was lovely that twenty-fifth anniversary dinner and Chuck, thanks for thinking of me today."

He was determined to have words with Mr. Bernstein. He was sure that he would understand. It was now or never. Come on Chuck, you can do it, he told himself.

Amid all the conversation Leslie suddenly screamed, dropped her champagne glass, and burst into tears. "O Chuck, Chuck," she sobbed, standing in the middle of the room over Chuck slumped in the recliner. A group of people stood around her trying to check her tears. One young man held her in his arms, patting her on the back.

Chuck had fallen asleep in his recliner, a deep sleep from which he was never going to wake.

ALEXANDER PUSHKIN

(RUSSIAN, 1799–1837)

From *Eugene Onegin*

45

The pail is brought, the ice is clinking
Round old Moët or Veuve Clicquot;*
This is what poets should be drinking
And they delight to see it flow.
Like Hippocrene it sparkles brightly,
The golden bubbles rising lightly
(The image, why, of this and that:
I quote myself, and do it pat).
I could not see it without gloating,
And once I gave my meagre all
To get it, friends, do you recall?
How many follies then were floating
Upon the magic of that stream—
What verse, what talk, how fair a dream!

*Producers of champagne.

46

But this bright sibilant potation
Betrays my stomach, and although
I love it still, at the dictation
Of prudence now I drink Bordeaux.
Aÿ* is risky, if delicious;
It's like a mistress, gay, capricious,
Enchanting, sparkling, frivolous,
And empty—so it seems to us . . .
But you, Bordeaux, I always treasure
As a good comrade, one who shares
Our sorrows and our smaller cares,
And also our calm hours of leisure,
One whose warm kindness has no end—
Long live Bordeaux, the faithful friend!

—Translated by Avraham Yarmolinsky

Y

HERBERT MORRIS

(A M E R I C A N , 1 9 2 8 –)

The Way Was Wine, the Voyage Was the Night

Some bouquets say Hemingway,
impetuous, a wind from Cuba,
domestics which suggest arrangements
scored for viola, flute, and tuba.

In the Médoc we came to claret
whose accent proved distinctly Chaucer,

---

*Champagne, referred to by its village in the Marne.

robust enough to warrant quaffing
not from a goblet but a saucer.

A Haut-Brion from Madagascar
spoke with the silences of Pinter,
proposing subtleties defying
even the vine, even the vintner.

I recall lyricism rampant,
a sauterne reeking of Verlaine,
so heady with defeat we sensed its
equal would not be poured again.

There was delirium of selfhood,
Lachrymae Christi on the Sound,
Rimbaud held in the Red Sea doldrums,
too soon on selfhood gone aground.

We sampled Pushkin on the Dnieper,
tender, severe, as mood decree.
We suffered ways by light, by water,
Nuits de St. Georges on the Black Sea.

There were the sagas fierce as retsin,
ravishments blond with sun, with Greece;
in the Aegean what consumed us
blinded more than the Golden Fleece.

One in Milan, Bologna, Parma,
Valpolicella wore the look
whose conflagrations lit the Titians,
whose rage in time we undertook.

There was Chignon, Chartreuse, Chenonceaux,
Baudelaire in the rancid streets,
the port of sailors aging, weaving,
gutted of hope, bereft of fleets.

Assailed by dusks, those turns to darkness,
mysteries of the night, ascensions,
we savored Hölderlin, but slowly,
derangement's infinite dimensions.

I remember a glass at Duino,
warbling of anguish over stones,
Rilke the chill that seared the castle,
what dread suffused the flesh, the bones.

We drank the shades against the evening,
the vin du pays from Aix to Ghent,
the bloom of verses steeped in acid,
young girls Apollinaire had sent.

There were the twilights turning darker,
dimness obscuring hallway, stair,
vapors and fumes of lamp-jets hissing,
whores for a thin vin ordinaire.

There was that France whose mists were morning,
vowels of light scooped from the bay,
cypresses drunk on moonlight's silver,
chablis that whispered Mallarmé.

We swallowed myth, at dawn sniffed absence,
took Yeats, but in a trembling hand;
and where the grapes had Lorca on them
the taste ran blood, the straits turned sand.

The routes plied south through brandy, whisky,
champagne flirtations, byway's gin.
Who was it said that where the wine is
voyages, mariner, begin?

Y

WILLIAM FAHEY

(AMERICAN, 1923-)

Marsala

Velvety brown and bitter-sweet
a thimbleful cupped on the tongue
swelling in the mouth like the nipple of Pomona*

Y

CHARLES BAUDELAIRE

(1821-1867)

Lovers' Wine

Look how splendid space is today!
Discard bit, spurs, bridle—away,
Let us saddle this horse like wine
And ride to a fairyland divine.

Like two angels tortured with sea-
Fever we will escape and flee,
Rise up and away to pursue
Morning's mirage of crystal blue.

Tenderly swaying on the wing
Of a quick whirlwind, in parallel
Desire, we ascend and sing,

* Roman goddess of fruit.

My sister, side by side, we shall
Reach out and swim in gentle streams
To the paradise of my dreams.

—Translated by John Digby

❦

MATTEOS ZARIFIAN

(A R M E N I A N , 1 8 9 4 – 1 9 2 4)

Ivresse

Fill the glass and let me drink
to purposeless delight
that mocks the sacred and profane of life.

Fill the glass and let me taste
the magic journey, wine,
that finds a road to heaven
through any pain.

Fill the glass and let me drink
the pungent fires of hell
while Death himself toasts in return
to my Good Health!

—Translated by Diana Der Hovanessian
and Marzbed Margossian

LIONEL JOHNSON

(ENGLISH, 1867–1902)

*Vinum Daemonum**

To Stephen Phillips

The crystal flame, the ruby flame,
Alluring, dancing, revelling!
See them: and ask me not, whence came
   This cup I bring.

But only watch the wild wine glow,
But only taste its fragrance: then,
Drink the wild drink I bring, and so
   Reign among men.

Only one sting, and then but joy:
One pang of fire, and thou art free.
Then, what thou wilt, thou canst destroy:
   Save only me!

Triumph in tumult of thy lust:
Wanton in passion of thy will:
Cry *Peace!* to conscience, and it must
   At last be still.

I am the Prince of this World: I
Command the flames, command the fires.
Mine are the draughts, that satisfy
   This World's desires.

*The spirit of wine.

Thy longing leans across the brink:
Ah, the brave thirst within thine eyes!
For there is that within this drink,
    Which never dies.

Y

JOHN STEINBECK

(A M E R I C A N , 1 9 0 2 – 1 9 6 8)

From *Tortilla Flat*

Two gallons is a great deal of wine, even for two paisanos. Spiritually the jugs may be graduated thus: Just below the shoulder of the first bottle, serious and concentrated conversation. Two inches farther down, sweetly sad memory. Three inches more, thoughts of old and satisfactory loves. An inch, thoughts of old and bitter loves. Bottom of the first jug, general and undirected sadness. Shoulder of the second jug, black, unholy despondency. Two fingers down, a song of death or longing. A thumb, every other song each one knows. The graduations stop here, for the trail splits and there is no certainty. From this point on anything can happen.

CHAPTER THREE
HERE'S TO HOPS

♟

TOBIAS SMOLLETT

(ENGLISH, 1721–1771)

From *Humphrey Clinker*

Well, there is no nation that drinks so hoggishly as the English—
What passes for wine among us, is not the juice of the grape. It is
an adulterous mixture, brewed up of nauseous ingredients, by
dunces, who are bunglers in the art of poison-making; and yet we,
and our forefathers, are and have been poisoned by this cursed
drench, without taste or flavour—The only genuine and wholesome beveridge in England, is London porter, and Dorchester ta-
ble-beer. . . .

♟

JOHN GAY

(1685–1732)

A Ballad on Ale

1
Whilst some in Epic strains delight,
Whilst others Pastorals invite,
    As taste or whim prevail;
Assist me, all ye tuneful Nine,
Support me in the great design,
    To sing of happy Ale.

2

Some folks of Cyder* make a rout,
Aud Cyder's well enough, no doubt,
   When better liquors fail;
But Wine, that's richer, better still,
Ev'n Wine itself (deny't who will)
   Must yield to nappy Ale.

3

Rum, Brandy, Gin with choicest smack
From *Holland* brought, *Batavia Arrack,*
   All these will nought avail
To chear a truly *British* heart,
And lively spirits to impart,
   Like humming, nappy Ale.

4

Oh! whether thee I closely hug
In honest can, or nut-brown jug,
   Or in the tankard hail;
In barrel, or in bottle pent,
I give the gen'rous spirit vent,
   Still may I feast on Ale.

5

But chief, when to the chearful glass
From vessel pure thy streamlets pass
   Then most thy charms prevail;
Then, then, I'll bett, and take odds,
That nectar, drink of heathen gods,
   Was poor, compar'd to Ale.

6

Give me a bumper, fill it up.
See how it sparkles in the cup,
   Oh how shall I regale!

*Cider is the traditional drink of Gay's native region, West England.

Can any taste this drink divine,
And then compare Rum, Brandy, Wine,
   Or aught with nappy Ale?

7

Inspir'd by thee, the warrior fights,
The lover wooes, the poet writes,
   And pens the pleasing tale;
And still in *Britain*'s isle confess'd
Nought animates the patriot's breast
   Like gen'rous, nappy Ale.

8

High Church and Low oft raise a strife,
And oft endanger limb and life,
   Each studious to prevail;
Yet *Whig* and *Tory* opposite*
In all things else, do both unite
   In praise of nappy Ale.

9

Inspir'd by thee shall *Crispin* sing,†
Or talk of freedom, church, and king,
   And balance *Europe*'s scale;
While his rich landlord lays out schemes
Of wealth, in golden *South Sea*‡ dreams,
   Th' effects of nappy Ale.

10

A blest potation! still by thee,
And thy companion Liberty,
   Do health and mirth prevail;
Then let us crown the can, the glass,
And sportive bid the minutes pass
   In quaffing nappy Ale.

*British political parties.
†Patron saint of shoemakers; hence he sings with a loose tongue.
‡A 1720 speculation hoax that ended with the collapse of South Sea Company stock and John Gay losing a lot of money.

11

Ev'n while these stanzas I indite,
The bar-bell's grateful sounds invite
    Where joy can never fail!
Adieu! my Muse, adieu! I haste
To gratify my longing taste
    With copious draughts of ALE.

Y

ANONYMOUS

(ENGLISH, SEVENTEENTH
CENTURY)

Wassail Song

Wassail!* wassail! all over the town,
Our bread is white, and our ale it is brown:
Our bowl it is made of the maplin tree,
So here, my good fellow, I'll drink to thee.

The wassailing bowl, with a toast within,
Come fill it up unto the brim;
Come fill it up, so that we may all see;
With the wassailing bowl I'll drink to thee.

Come, butler, come bring us a bowl of your best,
And we hope your soul in Heaven will rest;
But if you do bring us a bowl of your small,
Then down shall go butler, the bowl and all.

Oh, butler! oh, butler! now don't you be worst,
But pull out your knife and cut us a toast;

---

*From the Old English *Waes hael,* "to your health"; generally a spiced-ale drink associated with Christmas, New Year's, and Twelfth Night festivities. Wassail also refers to the carols sung on those occasions.

And cut us a toast, one that we may all see;—
With the wassailing bowl I'll drink to thee.

Here's to Dobbin, and to his right eye,
God send our mistress a good Christmas pye;
A good Christmas pye, as e'er we did see;—
With the wassailing bowl I'll drink to thee.

Here's to Broad May and to his broad horn,
God send our master a good crop of corn;
A good crop of corn, as we may all see,—
With the wassailing bowl I'll drink to thee.

Here's to Colly, and to her long tail,
We hope our master and mistress's heart will ne'er fail,
But bring us a bowl of your good strong beer,
And then we shall taste of your happy new year.

Be there here any pretty maids? we hope there be some,
Don't let the jolly wassailers stand on the cold stone,
But open the door, and pull out the pin,
That we jolly wassailers may all sail in.

JOHN CLARE

(E N G L I S H , 1 7 9 3 – 1 8 6 4)

The Toper's Rant *

Give me an old crone of a fellow
    Who loves to drink ale in a horn,
And sing racy songs when he's mellow,
    Which topers sung ere he was born.

* Heavy drinker.

For such a friend fate shall be thankèd,
    And, line but our pockets with brass,
We'd sooner suck ale through a blanket
    Than thimbles of wine from a glass.

Away with your proud thimble-glasses
    Of wine foreign nations supply,
A toper ne'er drinks to the lasses
    O'er a draught scarce enough for a fly.
Club me with the hedger and ditcher
    Or beggar that makes his own horn,
To join o'er an old gallon pitcher
    Foaming o'er with the essence of corn.

I care not with whom I get tipsy
    Or where with brown stout I regale,
I'll weather the storm with a gipsy
    If he be a lover of ale.

I'll weather the toughest storm weary
    Altho' I get wet to the skin,
For my outside I never need fear me
    While warm with real stingo* within.
We'll sit till the bushes are dropping
    Like the spout of a watering pan,
And till the cag's† drained there's no stopping,
    We'll keep up the ring to a man.
We'll sit till Dame Nature is feeling
    The breath of our stingo so warm,
And bushes and trees begin reeling
    In our eyes like to ships in a storm.

We'll start it three hours before seven,
    When larks wake the morning to dance,
And we'll stand it till night's black eleven,
    When witches ride over to France;

* Strong ale.
† Keg.

And we'll sit it in spite of the weather
   Till we tumble dead drunk on the plain,
When the morning shall find us together,
   All willing to stand it again.

♥

FRANCIS FAWKES

(E N G L I S H , 1 7 2 0 – 1 7 7 7)

The Brown Jug: A Song
Imitated from the Latin of Hieronymus Amaltheus

Dear Tom, this brown jug that now foams with mild ale,
(In which I will drink to sweet Nan of the Vale)
Was once Toby Fillpot,* a thirsty old soul
As e'er drank a bottle, or fathom'd a bowl;
In boosing about 'twas his praise to excel,
And among jolly topers he bore off the bell.

It chanc'd as in dog-days he sat at his ease
In his flow'r-woven arbour as gay as you please,
With a friend and a pipe puffing sorrows away,
And with honest old stingo was soaking his clay,
His breath-doors of life on a sudden were shut,
And he died full as big as a Dorchester butt.

His body, when long in the ground it had lain,
And time into clay had resolv'd it again,
A potter found out in its covert so snug,
And with part of fat Toby he form'd this brown jug,
Now sacred to friendship, and mirth, and mild ale,
So here's to my lovely sweet Nan of the Vale.

*The poem tells fancifully the origin of the Toby jug, a mug in the shape of a stout man wearing a long coat and a three-cornered hat.

Y

W. H. DAVIES

(ENGLISH, 1871–1940)

To Bacchus

I'm none of those—Oh Bacchus, blush!
   That eat sour pickles with their beer,
To keep their brains and bellies cold;
   Ashamed to let one laughing tear
Escape their hold.

For only just to smell your hops
   Can make me fat and laugh all day,
With appetite for bread and meat:
   I'll not despise bruised apples, they
Make cider sweet.

'Tis true I only eat to live,
   But how I live to drink is clear;
A little isle of meat and bread,
   In one vast sea of foaming beer,
And I'm well fed.

Ale

Now do I hear thee weep and groan,
   Who hast a comrade sunk at sea?
Then quaff thee of my good old ale,
   And it will raise him up for thee;
Thou'lt think as little of him then
As when he moved with living men.

If thou hast hopes to move the world,
    And every effort it doth fail,
Then to thy side call Jack and Jim,
    And bid them drink with thee good ale;
So may the world, that would not hear,
Perish in hell with all your care.

One quart of good old ale, and I
    Feel then what life immortal is:
The brain is empty of all thought,
    The heart is brimming o'er with bliss;
Time's first child, Life, doth live; but Death,
The second, hath not yet his breath.

Give me a quart of good old ale,
    Am I a homeless man on earth?
Nay, I want not your roof and quilt,
    I'll lie warm at the moon's cold hearth,
No grumbling ghost to grudge my bed,
His grave, ha! ha! holds up my head.

Y

HILAIRE BELLOC

(ENGLISH, BORN FRANCE,
1870–1953)

West Sussex Drinking Song

They sell good Beer at Haslemere
    And under Guildford Hill.
At Little Cowfold as I've been told,
    A beggar may drink his fill:
There is a good brew in Amberley too,
    And by the bridge also;

But the swipes they take in at Washington Inn
  Is the very best Beer I know.

Chorus

With my here it goes, there it goes,
  All the fun's before us:
The Tipple's* Aboard and the night is young,
The door's ajar and the Barrel is sprung,
I am singing the best song ever was sung
  And it has a rousing chorus.

If I were what I never can be,
  The master or the squire:
If you gave me the hundred from here to the sea,
  Which is more than I desire:
Then all my crops should be barley and hops,
  And did my harvest fail
I'd sell every rood of mine acres I would
  For a belly-full of good Ale.

Chorus

With my here it goes, there it goes,
  All the fun's before us:
The Tipple's aboard and the night is young,
The door's ajar and the Barrel is sprung,
I am singing the best song ever was sung
  And it has a rousing Chorus.

Drinking Dirge

A thousand years ago I used to dine
  In houses where they gave me such regale
Of dear companionship and comrades fine
  That out I went alone beyond the pale;

---

*Drink. The same word may be used to refer to the tapster and the drinker.

And riding, laughed and dared the skies malign
    To show me all the undiscovered tale—
But my philosophy's no more divine,
    I put my pleasure in a pint of ale.

And you, my friends, oh! pleasant friends of mine,
    Who leave me now alone, without avail,
On California hills you gave me wine,
    You gave me cider-drink in Longuevaille; *
If after many years you come to pine
    For comradeship that is an ancient tale—
You'll find me drinking beer in Dead Man's Chine.
    I put my pleasure in a pint of ale.

In many a briny boat I've tried the brine,
    From many a hidden harbour I've set sail,
Steering towards the sunset where there shine
    The distant amethystine islands pale.
There are no ports beyond the far sea-line,
    Nor any halloa to meet the mariner's hail;
I stand at home and slip the anchor-line.
    I put my pleasure in a pint of ale.

Envoi
Prince! Is it true that when you go to dine
    You bring your bottle in a freezing pail?
Why then you cannot be a friend of mine.
    I put my pleasure in a pint of ale.

*A village in north central France.

AHARON DADOVRIAN

(ARMENIAN, 1887–1965)

Czech Beer

The sun is down. Come, let us drink
while I tell you this beer's tale.
We think we lift its breathing light
against our lips. Instead we are its prey.

There was once, a wild-haired girl,
mad with love, who ran up into the hills
collecting wood to make a fire,
to match her wild and burning heart.

She wanted enough wood to start
a bonfire big enough to burn
all the love inside her heart:
"Let the wind take me to my beloved."

A sweet talking brook running by
heard her and shouted, Stop!
Crazy girl, whoever burns love
is punished by love. I know

where your love hides.
I'll make a bargain.
If you give me your blond hair
I'll bring him to you until morning.

Ecstatic, the girl agreed at once
and plucked out every last
golden thread
and gave it to the lustful brook,

who lied, and stole, who cheated her
ran off with the golden froth
and left her there
while it became Czech beer.

The sun is up! Let's drink again
to the woman here in the cup.
You can become her beloved now.
To woman, woman, woman. Drink.

—*Translated by Diana Der Hovanessian*

GUY DE MAUPASSANT

(F R E N C H , 1 8 5 0 – 1 8 9 3)

Waiter, a Bock!

Why on this particular evening, did I enter a certain beer shop? I cannot explain it. It was bitterly cold: A fine rain, a watery mist floated about, veiling the gas jets in a transparent fog, making the pavements under the shadow of the shop fronts glitter, which revealed the soft slush and the soiled feet of the passers-by.

I was going nowhere in particular; was simply having a short walk after dinner. I had passed the Credit Lyonnais, the Rue Vivienne, and several other streets. Suddenly I descried a large *café*, which was more than half full. I walked inside, with no object in mind. I was not the least thirsty.

By a searching glance I detected a place where I would not be too much crowded. So I went and sat down by the side of a man who seemed to me to be old, and who smoked a half-penny clay pipe, which had become as black as coal. From six to eight beer saucers were piled up on the table in front of him, indicating the number of "bocks" he had already absorbed. With that same glance

I had recognized in him a "regular toper," one of those frequenters of beer-houses, who come in the morning as soon as the place is open, and only go away in the evening when it is about to close. He was dirty, bald to about the middle of the cranium, while his long gray hair fell over the neck of his frock coat. His clothes, much too large for him, appeared to have been made for him at a time when he was very stout. One could guess that his pantaloons were not held up by braces, and that this man could not take ten paces without having to pull them up and readjust them. Did he wear a vest? The mere thought of his boots and the feet they enveloped filled me with horror. The frayed cuffs were as black at the edges as were his nails.

As soon as I had sat down near him, this queer creature said to me in a tranquil tone of voice:

"How goes it with you?"

I turned sharply round to him and closely scanned his features, whereupon he continued:

"I see you do not recognize me."

"No, I do not."

"Des Barrets."

I was stupefied. It was Count Jean des Barrets, my old college chum.

I seized him by the hand, so dumfounded that I could find nothing to say. I, at length, managed to stammer out:

"And you, how goes it with you?"

He responded placidly:

"With me? Just as I like."

He became silent. I wanted to be friendly, and I selected this phrase:

"What are you doing now?"

"You see what I am doing," he answered, quite resignedly.

I felt my face getting red. I insisted:

"But every day?"

"Every day is alike to me," was his response, accompanied with a thick puff of tobacco smoke.

He then tapped on the top of the marble table with a sou, to attract the attention of the waiter, and called out:

"Waiter, two 'bocks.' "

A voice in the distance repeated:

"Two 'bocks,' instead of four."

Another voice, more distant still, shouted out:

"Here they are, sir, here they are."

Immediately there appeared a man with a white apron, carrying two "bocks," which he set down foaming on the table, the foam running over the edge, on to the sandy floor.

Des Barrets emptied his glass at a single draught and replaced it on the table, sucking in the drops of beer that had been left on his mustache. He next asked:

"What is there new?"

"I know of nothing new, worth mentioning, really," I stammered: "But nothing has grown old for me; I am a commercial man."

In an equable tone of voice, he said:

"Indeed—does that amuse you?"

"No, but what do you mean by that? Surely you must do something!"

"What do you mean by that?"

"I only mean, how do you pass your time!"

"What's the use of occupying myself with anything. For my part, I do nothing at all, as you see, never anything. When one has not got a sou one can understand why one has to go to work. What is the good of working? Do you work for yourself, or for others? If you work for yourself you do it for your own amusement, which is all right; if you work for others, you reap nothing but ingratitude."

Then sticking his pipe into his mouth, he called out anew:

"Waiter, a 'bock.' It makes me thirsty to keep calling so. I am not accustomed to that sort of thing. Yes, I do nothing; I let things slide, and I am growing old. In dying I shall have nothing to regret. If so, I should remember nothing, outside this public-house. I have no wife, no children, no cares, no sorrows, nothing. That is the very best thing that could happen to one."

He then emptied the glass which had been brought him, passed his tongue over his lips, and resumed his pipe.

I looked at him stupefied and asked him:

"But you have not always been like that?"

HERE'S TO HOPS · 139

"Pardon me, sir; ever since I left college."

"It is not a proper life to lead, my dear sir; it is simply horrible. Come, you must indeed have done something, you must have loved something, you must have friends."

"No; I get up at noon, I come here, I have my breakfast, I drink my 'bock'; I remain until evening, I have my dinner, I drink 'bock.' Then about one in the morning, I return to my couch, because the place closes up. And it is this latter that embitters me more than anything. For the last ten years, I have passed six-tenths of my time on this bench, in my corner; and the other four-tenths in my bed, never changing. I talk sometimes with the *habitués*." *

"But on arriving in Paris what did you do at first?"

I paid my *devoirs* † "to the Café de Medicis."

"What next?"

"Next? I crossed the water and came here."

"Why did you take even that trouble?"

"What do you mean? One cannot remain all one's life in the Latin Quarter. The students make too much noise. But I do not move about any longer. Waiter, a 'bock.' "

I now began to think that he was making fun of me, and I continued:

"Come now, be frank. You have been the victim of some great sorrow; despair in love, no doubt! It is easy to see that you are a man whom misfortune has hit hard. What age are you?"

"I am thirty years of age, but I look to be forty-five at least."

I looked him straight in the face. His shrunken figure, badly cared for, gave one the impression that he was an old man. On the summit of his cranium, a few long hairs shot straight up from a skin of doubtful cleanness. He had enormous eyelashes, a large mustache, and a thick beard. Suddenly I had a kind of vision, I know not why—the vision of a basin filled with noisome water, the water which should have been applied to that poll. I said to him:

"Verily, you look to be more than that age. Of a certainty you must have experienced some great disappointment."

He replied:

"I tell you that I have not. I am old because I never take air.

* Regulars.
† Respects.

There is nothing that vitiates the live of a man more than the atmosphere of a *café*."

I could not believe him.

"You must surely have been married as well? One could not get baldheaded as you are without having been much in love."

He shook his head, sending down his back little hairs from the scalp:

"No, I have always been virtuous."

And raising his eyes toward the luster, which beat down on our heads, he said:

"If I am baldheaded, it is the fault of the gas. It is the enemy of hair. Waiter, a 'bock.' You must be thirsty also?"

"No, thank you. But you certainly interest me. When did you have your first discouragement? Your life is not normal, is not natural. There is something under it all."

"Yes, and it dates from my infancy. I received a heavy blow when I was very young. It turned my life into darkness, which will last to the end."

"How did it come about?"

"You wish to know about it? Well, then, listen. You recall, of course, the castle in which I was brought up, seeing that you used to visit it for five or six months during the vacations? You remember that large, gray building in the middle of a great park, and the long avenues of oaks, which opened toward the four cardinal points! You remember my father and my mother, both of whom were ceremonious, solemn, and severe.

"I worshiped my mother; I was suspicious of my father; but I respected both, accustomed always as I was to see everyone bow before them. In the country, they were Monsieur le Comte and Madame la Comtesse; and our neighbors, the Tannemares, the Ravelets, the Brennevilles, showed the utmost consideration for them.

"I was then thirteen years old, happy, satisfied with everything, as one is at that age, and full of joy and vivacity.

"Now toward the end of September, a few days before entering the Lycée, while I was enjoying myself in the mazes of the park, climbing the trees and swinging on the branches, I saw crossing an avenue my father and mother, who were walking together.

"I recall the thing as though it were yesterday. It was a very windy day. The whole line of trees bent under the pressure of the wind, moaned and seemed to utter cries—cries dull, yet deep—so that the whole forest groaned under the gale.

"Evening had come on, and it was dark in the thickets. The agitation of the wind and the branches excited me, made me skip about like an idiot, and howl in imitation of the wolves.

"As soon as I perceived my parents, I crept furtively toward them, under the branches, in order to surprise them, as though I had been a veritable wolf. But suddenly seized with fear, I stopped a few paces from them. My father, a prey to the most violent passion, cried:

" 'Your mother is a fool; moreover, it is not your mother that is the question, it is you. I tell you that I want money, and I will make you sign this.'

"My mother responded in a firm voice:

" 'I will not sign it. It is Jean's fortune, I shall guard it for him and I will not allow you to devour it with strange women, as you have your own heritage.'

"Then my father, full of rage, wheeled round and seized his wife by the throat, and began to slap her full in the face with the disengaged hand.

"My mother's hat fell off, her hair became disheveled and fell down her back: she essayed to parry the blows, but could not escape from them. And my father, like a madman, banged and banged at her. My mother rolled over on the ground, covering her face in both her hands. Then he turned her over on her back in order to batter her still more, pulling away the hands which were covering her face.

"As for me, my friend, it seemed as though the world had come to an end, that the eternal laws had changed. I experienced the overwhelming dread that one has in presence of things supernatural, in presence of irreparable disaster. My boyish head whirled round and soared. I began to cry with all my might, without knowing why, a prey to terror, to grief, to a dreadful bewilderment. My father heard me. I believed that he wanted to kill me, and I fled like a hunted animal, running straight in front of me through the woods.

"I ran perhaps for an hour, perhaps for two, I know not. Darkness had set in, I tumbled over some thick herbs, exhausted, and I lay there lost, devoured by terror, eaten up by a sorrow capable of breaking forever the heart of a child. I became cold, I became hungry. At length day broke. I dared neither get up, walk, return home, nor save myself, fearing to encounter my father whom I did not wish to see again.

"I should probably have died of misery and of hunger at the foot of a tree if the guard had not discovered me and led me by force.

"I found my parents wearing their ordinary aspect. My mother alone spoke to me:

" 'How you have frightened me, you naughty boy; I have been the whole night sleepless.'

"I did not answer, but began to weep. My father did not utter a single word.

"Eight days later I entered Lycée.

"Well, my friend, it was all over with me. I had witnessed the other side of things, the bad side; I have not been able to perceive the good side since that day. What things have passed in my mind, what strange phenomena have warped my ideas, I do not know. But I no longer have a taste for anything, a wish for anything, a love for anybody, a desire for anything whatever, no ambition, no hope. And I always see my poor mother lying on the ground, in the avenue, while my father was maltreating her. My mother died a few years after; my father lives still. I have not seen him since. Waiter, a 'bock.' "

A waiter brought him his "bock," which he swallowed at a gulp. But, in taking up his pipe again, trembling as he was, he broke it. Then he made a violent gesture:

"Zounds! This is indeed a grief, a real grief. I have had it for a month, and it was coloring so beautifully!"

Then he went off through the vast saloon, which was now full of smoke and of people drinking, calling out:

"Waiter, a 'bock'—and a new pipe."

ROBERT MINHINNICK

(ANGLO-WELSH, 1952-)

The Drinking Art

The altar of glasses behind the bar
Diminishes our talk. As if in church
The solitary men who come here
Slide to the edges of each black
Polished bench and stare at their hands.
  The landlord keeps his own counsel.

This window shows a rose and anchor
Like a sailor's tattoo embellished
In stained glass, allows only the vaguest
Illumination of floor and ceiling,
The tawny froth the pumps sometimes spew.
  And the silence settles. The silence settles

Like the yellow pinpoints of yeast
Falling through my beer, the bitter
That has built the redbrick
Into the faces of these few customers,
Lonely practitioners of the drinking art.
  Ashtrays, a slop-bucket, the fetid

Shed-urinal, all this I wondered at,
Running errands to the back-doors of pubs,
Woodbines and empty bottles in my hands.
Never become a drinking-man, my
Grandmother warned, remembering Merthyr*
  And the Spanish foundrymen

* An industrial-steel town in northeast Glamorgan, Wales.

Puking their guts up in the dirt streets,
The Irish running from the furnaces
To crowd their paymaster into a tavern,
Leather bags of sovereigns* bouncing on his thigh.
But it is calmer here, more subtly dangerous.
   This afternoon is a suspension of life

   I learn to enjoy. But now
The towel goes over the taps and I feel
The dregs in my throat. A truce has ended
And the clocks start again. Sunlight
Leaps out of the street. In his shrine of glass
   The landlord is wringing our lives dry.

Y

ANONYMOUS

(ENGLISH, SEVENTEENTH
CENTURY)

Song in Praise of Ale

Submit, bunch of grapes,
To the strong barley ear;
The weak wine no longer
The laurel shall wear.

Sack, and all drinks else,
Desist from the strife;
Ale's the only Aqua vitæ,
And liquor of life.

Then come, my boon fellows,
Let's drink it around;

*Gold coins worth one pound or twenty old shillings.

It keeps us from grave,
Though it lays us on ground.

Ale's a physician,
No mountebank bragger;
Can cure the chill ague,
Though it be with the stagger.

Ale's a strong wrestler,
Kings all it hath met:
And makes the ground slippery,
Though it be not wet.

Ale is both Ceres,
And good Neptune too,*
Ale's froth was the sea,
From which Venus grew.

Ale is immortal;
And be there no stops,
In bonny lads quaffing,
Can live without hops.

Then come, my boon fellows,
Let's drink it around;
It keeps us from grave,
Though it lays us on ground.

*The mythological deities of grain and water are equated with ale.

Y

JOHN DIGBY

(1 9 3 8 –)

A Soldier's Beer

All night the guns roared loud and clear.
We huddled like rats in the trench's mud;
I thought of England's cricket fields, and a beer
That would course and cruise through my blood.

All night the shells made a deafening noise
Hammering on our battered ears
And still I dreamt of that beer with the boys
In a pub back home beyond war's frontiers.

"Retreat," the sergeant screamed overhead.
I turned to tap my mate on the shoulder.
"Forget him," the sergeant said, "he's dead."
We scrambled up as the enemy came on bolder.

We sheltered in a wood: our sergeant spoke,
"Old Tommy was dreaming of England and a beer—
We'll bury him later, wasn't such a bad bloke.
Never dream my lads," his lip curled a sneer.

"If you dream you'll drift and forget
There's a war going on out there at night."
The sergeant uttered his words as a threat,
The dream of my beer shattered and took flight.

♀

JAMES STEPHENS

(IRISH, 1882–1950)

A Glass of Beer

The lanky hank of a she in the inn over there
Nearly killed me for asking the loan of a glass of beer;
May the devil grip the whey-faced slut by the hair,
And beat bad manners out of her skin for a year.

That parboiled ape, with the toughest jaw you will see
On virtue's path, and a voice that would rasp the dead,
Came roaring and raging the minute she looked at me,
And threw me out of the house on the back of my head!

If I asked her master he'd give me a cask a day;
But she, with the beer at hand, not a gill would arrange!
May she marry a ghost and bear him a kitten, and may
The High King of Glory permit her to get the mange.

♀

JOÃO GUIMARÃES ROSA

(BRAZILIAN, 1908–1967)

The Horse That Drank Beer

The man's country house was darkened and half-hidden by the
trees; you never saw so many trees around a house. He was a
foreigner. My mother told me how he came the year of the Span-

ish flu, cautious and scared-like, to buy that place that was so easy to defend. From any of the windows of the house you could keep watch for a long distance, with your hand on your shotgun. In those days, he still hadn't grown so fat it made you sick. They said he ate all kinds of dirt: snails, even frogs, with armfuls of lettuce soaked in a bucket of water. You see, he ate lunch and supper sitting outside on the stoop, the bucket with the lettuce in it on the ground between his thick legs. The meat he cooked, though; that was real beef off a cow. What he spent most of his money on was beer, but he didn't drink it in front of folks. I used to pass by and he'd ask me: "Irivalini, *bisogno** another bottle, it's for the horse." I don't like to ask questions, and I didn't think it was funny. Sometimes I didn't bring it; but sometimes I did, and then he'd pay me back and give me a tip. Everything about him made me mad. He couldn't even learn to say my name right. Insult or just disrespect, I'm not the kind to forgive any mother's son either way.

Mother and I and a few other people used to pass by his gate on the way to the plank bridge over the creek. "Let him be, poor thing, he suffered in the war," my mother would say. There were always a lot of dogs around him, big fellows, guarding the house. One of them he didn't like. You could see the beast was scared and unsociable—it was the one he treated the worst—but even so, it wouldn't budge from his side, and though he despised it, he was always calling the poor devil. Its name was Mussulino.

Well, I brooded over the grudge I had against him, a bull-necked, paunchy man like that, hoarse and snuffling all the time, so foreign it turned your stomach—was it fair for him to have all that money and position, to come and buy Christian land with no respect for people who were hard up, to order dozens of beers and not even pronounce the word right? Beer? The fact was, he did have some horses, four or maybe three, and they sure led an easy life. He never rode any of them, couldn't have mounted them anyway. Why, he could hardly even walk. The old billy goat! He never stopped puffing on his little smelly cigar, all chewed up and drooled on. He had a good beating coming to him. So damned careful with his house locked up, like he thought everybody else was a thief.

*I need.

He did admire my mother and treated her nice and kind. With me, though, it was no use—I couldn't help hating him, even when Mother was so sick and he offered money to pay for the medicine. I took it; who can live off noes? But I didn't thank him. He probably felt guilty for being a rich foreigner. And it didn't do any good anyway; my mother, a real saint, went off into the dark, and damned if he didn't want to pay for the funeral. Later on he asked me if I'd like to come and work for him. I turned it over in my mind and argued with myself about what his reasons might be. He knew I was brave, that I had my pride, and that I could face up to anything; not many people in those parts were willing to look me straight in the eye. I thought it must be because he wanted my protection, day and night, against undesirable strangers and such. It must have been that, because he gave me hardly anything to do; I could fool around all I wanted to, so long as I had a weapon close by. But I did go to the store for him. "Beer, Iriva-lini. For the horse," was what he'd say, with a long face, in that language you could beat eggs with. I wished he'd cuss me! Then he'd see what was what.

The thing I couldn't get used to was all that covering up. That big old house was kept locked night and day; no one ever went inside—not even to eat, not even to cook. Everything went on out of doors. I reckon he hardly ever went in there himself, except to sleep or put away the beer—haw, haw—the beer for the horse. And I said to myself: "You wait, you pig, and see if one of these days I don't get in there, no matter what!" Maybe that was when I should have talked to the right people, told them what crazy things were going on, asked them to do something about it, put a bug in their ear. But it was easier not to. Words don't come easy to me. And then, just about that time, they came—the strangers.

Two jokers from the capital. It was Seo Priscílio, the deputy police commissioner, who called me over to them. He said: "Rei-valino Simple, these men you see here are officials; they're on the up and up." And then the strangers took me aside and tried to pump me. They wanted to know all about the man's habits, and they asked me all sorts of silly questions. I put up with it, but I didn't give anything away. What am I, anyway, a coati for dogs to bark at? I had my doubts too when I looked at the ugly mugs of those bundled-up rascals. But they paid me; a fair sum of money, too. The leader of the two, rubbing his chin with his hand, gave me a job to do: to find out whether my boss, "a very dangerous man," really lived by himself. And I was to take a good look, the first chance I got, and see if he didn't have, low down on his leg, the old mark of a collar, an iron ring, the kind criminals wear. And me, I piped up and promised to do it.

Dangerous, to me? Ha, ha. Maybe he had been a real man when he was young. But now he was paunchy, a high liver, lazy as they come, and all he wanted was beer—for the horse. The rat. Not that I had any complaints; I never liked beer much. If I had, I would have bought some for myself, or drunk his, or asked him for some; he would have given it to me. He said he didn't like it, either. And come to think of it, I guess he didn't. All he ever ate was that pile of lettuce with meat, his mouth so full it'd make you sick, along with a lot of olive oil, and he'd lick the drippings. He'd been kind of angry and suspicious lately; did he know about the strangers' visit? I didn't see any slave brand on his leg; didn't even try to look. Am I some kind of errand boy to a chief deputy, one of those heavy thinkers who's always signing papers? But I did want to have a peek at that locked-up house, even if it had to be through a crack. And now that the dogs were tame and friendly, it seemed as if I could manage it. But Seo Giovânio got the idea that I was up to something, because one day he gave me a surprise: he called me over and opened the door. It stank inside, the way things do when the top's never taken off; the air didn't smell right. The parlor was big and empty; no furniture, just space. He let me look all I wanted to, on purpose it seemed, walked with me through several rooms, till I was satisfied. Ah, but later on I talked it over with myself, and I had an idea: what about the bedrooms? There were a lot of them left over; I hadn't gone into all of them by a long shot. Behind one of those doors I could feel something breathing—or did I only think that later? Oh, so the wop wanted to cheat me. Didn't he know I was smarter than he was?

Well, after a few days had gone by, very late on two or three different nights you could hear galloping out on the empty plain, like a horseman riding out at the pasture gate. Was the fellow fooling me, turning himself into a ghost or a werewolf? On the other hand, there was that trick his mind had of wandering off the track in a way I could never quite understand; that might explain part of it. What if he really kept some funny kind of horse always hidden there inside, in the dark house?

Seo Priscílio called on me one more time that week. The strangers were there with him on the quiet. I came in in the middle of the conversation: I made out that one of those two worked for the "Consulate." But I told them everything, or a lot anyway, to get even. Then the strangers got Seo Priscílio all fired up to carry out their plan. They wanted to stay under cover and let him go alone. And they gave me some more money.

I hung around pretending to be deaf and dumb, just twiddling my thumbs. Seo Priscílio came and spoke to Seo Giovânio: what were those tales he had been hearing about a horse that drank beer? He kept on pressing him, trying to learn something. Seo Giovânio looked awfully tired. He shook his head slow and weary and snuffled the snot from his nose to the stump of his cigar; but he didn't get mad at him. He kept putting his hand to his forehead: "You wanta see?" Then he went off and came back with a basket with lots of full bottles and a pail and poured it all in, with the beer foaming up. He told me to go get the horse: the light cinnamon-colored sorrel with the pretty face. The animal trotted up raring to go—would you believe it?—with his ears twitching and his nostrils flared and his tongue hanging out, and smackingly drank down the beer. He enjoyed it clear down to the bottom of the pail. You could see he was used to drinking all he wanted! When had he been taught that trick, of all things? Well, that horse just couldn't drink enough. Seo Priscílio looked ashamed of himself, thanked Seo Giovânio, and went away. My boss let out a whistle and looked at me: "Irivalini, times are getting bad. Don't leave the guns!" I approved of that even if I did have to laugh at his crazy ideas and his tall tales. I still halfway hated him.

So when the strangers came back again, I told them what I thought: that there must be something else in the bedrooms in that house. This time Seo Priscílio came with a soldier; he simply announced that he intended to search all the rooms in the name of the law! Seo Giovânio, standing still and peaceful-like, just lit another cigar; he was always cool and sensible. He opened up the house so that Seo Priscílio could go in, and the soldier, too, and me. The bedrooms? He went straight to one that was all bolted tight. And lo and behold! There inside, big as a house, was the queerest thing—was I dreaming?—a big white horse, a stuffed horse. It was so exactly right with its square face, like a hobby-horse; shiny-bright, white, clean, with a long mane and fat hindquarters, tall as the one in the church—a St. George horse. How could they have brought it, or had it brought, and got it in there all put together? Seo Priscílio's face fell; he couldn't get over it. He felt the horse all over, but it wasn't hollow; it just had straw inside. Seo Giovânio, as soon as he was alone with me, chewed on his cigar: "Irivalini, it's a shame the two of us don't like beer, eh?" It sure was. I felt like telling him what was behind all the goings-on.

But if Seo Priscílio and the two strangers weren't suspicious any more, what about the other bedrooms in the house, and behind the doors? They should have searched all over and been done with it. Not that I was going to remind them, not being the kind to pop my hand up to be called on all the time. Seo Giovânio started talking to me again, after he had mulled things over a while: "Iri-valini, *ecco,* life is hard and men are prisoners . . ." I didn't want to ask him about the white horse. Fiddlesticks. What did I care about it? It must have been his horse during the war; it was prob-ably his pet. "But Irivalini, we love life too much." He wanted me to eat with him, but his nose was dripping with snot and dribble, and he sniffled and didn't bother to wipe it off, and every inch of him stank of cigar smoke. It was awful to watch that man not feeling sorry for himself. So I went and talked to Seo Priscílio: said I wouldn't have anything more to do with those strangers who wanted to stir up trouble. I wouldn't play their two-faced game! If they came back I'd run them off the place, I'd cut loose, I'd fight them—just hold your horses! This is Brazil, and they were just as much foreigners as the boss was. I'm quick to draw a knife or a gun. Seo Priscílio knew it. But I don't think he knew about the surprise that was coming.

It happened all of a sudden. Seo Giovânio threw the house wide open. Then he called me: in the parlor, in the middle of the floor, was a man's corpse lying under a sheet. "Josepe, my brother," he said to me, looking all cut up. He wanted the priest, wanted the church bell to toll three times three, wanted the sad part to be just right. No one had ever known that brother of his who had run away from people and kept himself hidden. It was a respectable funeral. Seo Giovânio needn't feel ashamed before folks. Except that Seo Priscílio came beforehand—I reckon the strangers had promised him some money—and made us lift up the sheet for him to have a look. But all you could see was how holy-terrified we all were, and our staring eyes: the dead man had no face, you might say—only a great big gaping hole, with old healed-over scars, awful, no nose, no cheeks—you could see the white bones where his throat began, his gullet and his tonsils. "That's-a the war," explained Seo Giovânio, forgetting to close his silly mouth that was hanging open, all nasty sweetness.

Now I wanted to get out of there and go on my way; it was no good for me to stay there any longer in that crazy, unlucky place, with the trees so dark all around. Seo Giovânio was sitting outside, the way he had done for so many years. He looked sicker and older all of a sudden, with that sharp pain sticking into him. But he was eating his meat and the heads of lettuce from the bucket, and snuffling. "Irivalini . . . this life is . . . *bisogno. Capisce?"* * he asked in his singsong voice, looking at me, his face purple. "*Capisco,*" I answered. I didn't give him an *abraço;* † not that I was disgusted, just ashamed. I didn't want to cry, too. And then if he didn't do the damnedest thing: he opened the beer and watched it foaming out on the ground. "*Andiamo,* Irivalini, *contadino, bambino?"*‡ he proposed. I said yes. Glass after glass, by the twenties, the thirties, I drank it all down. He had calmed down some by then, and when I went away he asked me to take the horse—the sorrel that drank beer—and that hangdog thin dog, Mussulino.

I never saw my boss again. I knew when he died, because he left the farm to me in his will. I had tombstones made and masses said for him, and for his brother and my mother. I sold the place, but first I cut down all those trees and buried the stuff that had been in that room I told you about. I never went back there again. No, I can't forget that one day—how awful I felt. The two of us

*I need . . . you understand?
†An embrace.
‡Lets go . . . to the country, kid?

with all those many, many bottles. When it happened, I thought
I saw someone else come up suddenly behind us: the white-faced
sorrel, or St. George's big white horse, or the poor, miserable
brother. I, Reivalino Simple, *capiscoed*. I drank up all the beer that
was left, and now I pretend it was me that drank all the beer in
that house. Just so nobody will make any more mistakes.

—Translated by Barbara Shelby

Y

JOHN DIGBY

(1 9 3 8 –)

A Small Beer Complaint

Tell me barman where's the beer hidden?
I've had café in old Vienna,
I've known parts where liquor's forbidden,
Quenched my thirst with wine in Ravenna.
But where's the beer that knocks your block off?
Today, it's champagne, gin and brandy,
Madeira, rye, cognac and Smirnoff's
But right now a beer would be handy.

Tell me where are the beers of yesteryear?
Without beer life appears so austere.

I've made pub crawls from Iceland to Zanzibar,
Swilled back buckets of various liquors;
Most of them tasted like vinegar,
Though I've drunk with the most adroit kickers.
What puzzles me is: Where's the beer?
I've seen barmen shake cocktails till dizzy.
When I've asked for a beer, I've met with a sneer.
"We don't stock that. We're far too busy."

Tell me where are the beers of yesteryear?
Without beer I'm as lost as old King Lear.

Beer I swear is a most rare commodity.
I've known grown men to sit down and cry
When a brewery's taken a holiday.
I've known tragedies—I've seen pubs run dry!
It's all these concoctions that get to me:
Whiskey sour, marguerita and stinger,
Pink gin with a kick and Long Island Tea
But no beer that will make you a singer.

Tell me where are the beers of yesteryear?
Without beer I'm as dead as a stricken deer.

Outside the "Bald Faced Stag" I die of thirst
For want of a beer with a frothy head.
By the look of things I won't be the first!
I tell you thirst's a disease that can easily spread.

Tell me where are the beers of yesteryear?
Ask me not for beer's disappearing I fear.

CHAPTER FOUR

BLITHE SPIRITS

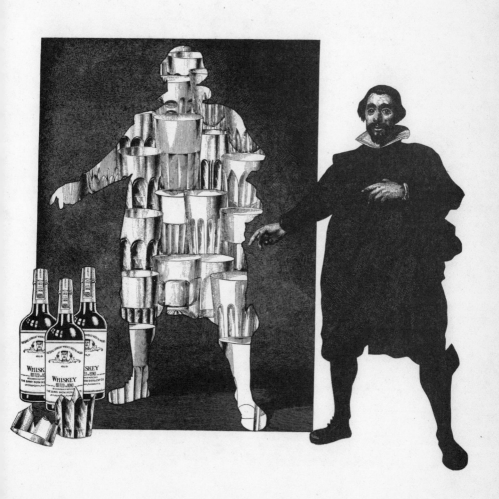

Y

SAMUEL HOFFENSTEIN

(AMERICAN, 1890–1947)

Hope that springs eternal in
The human breast, is fond of gin,
Or Scotch or beer or anything
Designed to help a hope to spring.

Y

LOUIS MACNEICE

(BRITISH, BORN NORTHERN
IRELAND, 1907–1963)

Alcohol

On golden seas of drink, so the Greek poet said,
Rich and poor are alike. Looking around in war
We watch the many who have returned to the dead
Ordering time-and-again the same-as-before:

Those Haves who cannot bear making a choice,
Those Have-nots who are bored with having nothing to choose,
Call for their drinks in the same tone of voice,
Find a factitious popular front in booze.

Another drink: Bacchylides* was right
And self-deception golden—Serve him quick,
The siphon stutters in the archaic night,
The flesh is willing and the soul is sick.

*Greek lyric poet, circa 500 B.C.

Another drink: Adam is back in the Garden.
Another drink: the snake is back on the tree.
Let your brain go soft, your arteries will harden;
If God's a peeping tom he'll see what he shall see.

Another drink: Cain has slain his brother.
Another drink: Cain, they say, is cursed.
Another and another and another—
The beautiful ideologies have burst.

A bottle swings on a string. The matt-grey iron ship,
Which ought to have been the Future, sidles by
And with due auspices descends the slip
Into an ocean where no auspices apply.

Take away your slogans; give us something to swallow,
Give us beer or brandy or schnapps or gin;
This is the only road for the self-betrayed to follow—
The last way out that leads not out but in.

Y

THOMAS LOVELL BEDDOES

(E N G L I S H , 1 8 0 3 – 1 8 4 9)

Lord Alcohol

1
Who tames the lion now?
Who smoothes Jove's wrinkles now?
Who is the reckless wight
    That in the horrid middle
Of the deserted night
Doth play upon man's brain,
    As on a wanton fiddle,

The mad and magic strain,
The reeling, tripping sound,
To which the world goes round?
     Sing heigh! ho! diddle!
          And then say—
Love, quotha, Love? nay, nay!
It is a spirit fine
Of ale or ancient wine,
     Lord Alcohol, the drunken fay,
          Lord Alcohol alway!

                    2
Who maketh the pipe-clay man
Think all that nature can?
Who dares the gods to flout,
     Lay fate beneath the table,
And maketh him stammer out
A thousand monstrous things,
     For history a fable,
Dish-clouts for kings?
And sends the world along
Singing a ribald song
     Of heigho! Babel?
          Who, I pray—
Love, quotha, Love? nay, nay!
It is a spirit fine
Of ale or ancient wine,
     Lord Alcohol, the drunken fay,
          Lord Alcohol alway!

Y

JOSEPH O'LEARY

(I R I S H , 1 7 9 0 – 1 8 5 0)

Whisky, Drink Divine

Whisky, drink divine!
  Why should drivelers bore us
With the praise of wine
  While we've thee before us?
Were it not a shame,
  Whilst we gayly fling thee
To our lips of flame,
  If we could not sing thee?

Chorus
Whisky, drink divine!
  Why should drivelers bore us
With the praise of wine
  While we've thee before us?

Greek and Roman sung
  Chian and Falernian—
Shall no harp be strung
  To thy praise, Hibernian?
Yes! let Erin's sons—
  Generous, brave, and frisky—
Tell the world at once
  They owe it to their whisky—

  Whisky, drink divine! *etc.*

If Anacreon—who
  Was the grape's best poet—

Drank our *mountain-dew,*
  How his verse would show it!
As the best then known,
  He to wine was civil;
He had *Inishowen,* *
  He'd pitch wine to the devil—

    Whisky, drink divine! *etc.*

Bright as beauty's eye,
  When no sorrow veils it:
Sweet as beauty's sign,
  When young love inhales it:
Come, then, to my lips—
  Come, thou rich in blisses!
Every drop I sip
  Seems a shower of kisses—

    Whisky, drink divine! *etc.*

Could my feeble lays
  Half thy virtues number,
A whole *grove* of bays
  Should my brows encumber.
Be his name adored,
  Who summed up thy merits
In one little word,
  When we call thee *spirits*—

    Whisky, drink divine! *etc.*

Send it gayly round—
  Life would be no pleasure,
If we had not found
  This enchanting treasure:
And when tyrant death's

*Whiskey.

Arrow shall transfix ye,
Let your latest breaths
Be whisky! whisky! whisky!

Whisky, drink divine! *etc.*

Y

ANTON CHEKHOV

(R U S S I A N , 1 8 6 0 – 1 9 0 4)

Carelessness

Pyotr Petrovitch Strizhin, the same whose rubbers had been stolen last year, was returning home from a Christening party at two o'clock in the morning.

In order not to arouse the household he quietly undressed in the hall and breathlessly tiptoed into his bedroom, where, without turning on a light, he was about to lie down.

Strizhin was a fellow with the face of a fool, living a regular and sober life and reading only books with a moral purpose. It was only on such a festive occasion as the present that, in honor of the newly born child of Lyubov Spirodovna, he allowed himself to empty four glasses of whiskey and a glass of wine which tasted like castor-oil. These hot beverages are like sea-water or fame: the more you quaff, the more thirsty you become. . . .

And now, while he was undressing, Strizhin suddenly acquired an overwhelming thirst.

"If I'm not mistaken," said he, "there is a bottle of whiskey in Dashenka's cupboard. It's right there in the corner, if I'm not mistaken. . . . She'll never notice the difference if I take only a little glass."

After a short deliberation he mastered his timidity and went to the cupboard.

Opening the door of the cupboard slowly, he found a bottle in the right-hand corner. He filled a glass in the dark, replaced the bottle, crossed himself, and swallowed the contents with a single gulp.

And here something remarkable happened. A terrible force, like a bomb, hurled Strizhin from the cupboard to the trunk. His eyes beheld a flash of lightning, he began to choke and shiver as though he had just fallen into a swamp full of blood-suckers. It seemed to him that instead of whiskey he had just swallowed a piece of dynamite which blew him to pieces and scattered his head, his arms, his legs, the house and the entire street in all directions, way up in the air, the devil knows whither. . . .

For about three minutes he lay motionless and breathless on the trunk. Then he rose and asked himself:

"Where am I?"

When he came to himself, he smelt for the first time a strong odor of kerosene.

"Holy Father in Heaven!" he cried, "I've drunk kerosene instead of whiskey!" And a shudder passed through his body.

The thought that he had poisoned himself threw him into an ague. And that he really had poisoned himself was evident not only from the odor in the room, but also from the sparks that danced before his eyes, from the ringing in his ears and from the stabbing pain in his stomach.

Realizing that he was about to die and unable to delude himself with false hopes, he decided to say good-bye to his nearest friends. He therefore went into Dashenka's bed-room, for he was a widower and Dashenka, an old maid, was keeping house for him.

"Dashenka," he said with a sob, "dear Dashenka!"

Something stirred in the dark and emitted a deep sigh.

"Dashenka!"

"Who is it?" said Dashenka with a start. "Oh, is it you, Pyotr Petrovitch? Are you back already? Well? Did they name the baby? Who was the god-mother?"

"Natalya Andreyevna was the godmother, and Pavel Ivanitch was the god-father, and I think I am dying, Dashenka. The baby was named Olympiada . . . and . . . I have drunk kerosene."

"What! You don't mean to say they gave you kerosene, do you?"

"I'll make a clean breast of it. I wanted to take a little nip of whiskey without getting your permission . . . and . . . God punished me for it. By mistake I took kerosene in the dark. . . . What shall I do now?"

When she heard that her cupboard had been opened without her permission she jumped to her feet. Hastily lighting a candle, she drew herself to her full angular and bony height, and forgetting to throw anything over her nightgown she shuffled in her bare feet over to the cupboard.

"Who gave you permission?" she cried, opening the cupboard. "Nobody ever put the whiskey there for your benefit!"

"I . . . I . . . have drunk kerosene, Dashenka, not whiskey . . ." mumbled Strizhin, wiping the cold perspiration from his face.

"And why should you be nosing around my kerosene? Is it any of your business? I didn't buy it for you, did I? If you only knew what a devilish price I've had to pay for it! You know nothing about that, of course *you* don't!"

"Dear Dashenka," he groaned, "it is a question of life and death, and you speak about money!"

"Not satisfied with getting drunk, he comes around sticking his nose into the cupboard!" raged Dashenka, shutting the door of the cupboard with a bang. "You bandit, you torturer, you! Not a minute of rest have I! By day and by night you ruin me! Robber, murderer! May you live in the next world as peacefully as you let me live here! Tomorrow I am going to leave this place. I am a virgin and I refuse to let you remain in my presence in your underwear! And don't you dare to look at me when I am undressed!"

She gave loose rein to her tongue and away it galloped, on and on and on. . . .

Knowing that nothing could be done with her in her fit of anger, that neither soft words nor harsh, neither prayers nor oaths, nor even bullets would be of any avail, Strizhin made a gesture of despair, dressed himself and went in search of a doctor. But doctors and policemen are found only when not needed. He ran through several streets, rang five times at one doctor's house, seven times at another's, and finally hastened over to the drug store. Perhaps the druggist would help.

After a long pause a little, dark, curly-headed druggist opened the door. He was dressed in his bathrobe and his eyes were still sleepy, but his face inspired a feeling of awe, so stern and intelligent was its expression.

"What can I do for you?" he asked in tones which only smart and worthy druggists know how to employ.

"For God's sake, I beseech you!" cried Strizhin breathlessly. "Give me something, anything! . . . I've just drunk kerosene! I'm dying!"

"Now, don't become excited, my good fellow, but answer the questions that I am going to put to you. Your nervous frame of mind makes it impossible for me to understand you. You have drunk kerosene. Am I right?"

"Yes! Kerosene! Save me!"

The druggist solemnly went to his counter, opened a book and was lost in deep meditation. Having perused a couple of pages, he shrugged one shoulder, then the other, made a grimace, mused for a minute or two and then went into a rear room. At this moment the clock struck four. At precisely a quarter to five he returned from the rear room with another book and once more was lost in deep perusal.

"H'm!" he finally said in perplexity. "The very fact that you feel bad shows that you ought to go to a doctor instead of a druggist."

"But I've already been to doctors! I rang and rang, but nobody answered."

"H'm! Evidently you don't regard us druggists as human beings. It's nothing to you to wake us up even at four o'clock at night. Every dog, every cat has its rest . . . But you will not listen to it. You imagine that we are not human, and that our nerves are made of hemp!"

Strizhin listened patiently till the druggist finished his harangue, sighed and went home.

"I guess it is my fate to die!" he thought.

His mouth was as hot as a furnace, his throat was choking with the odor of kerosene, his stomach writhed with cramps and his ears were deafened with a ringing noise: boom! boom! boom! Every moment he thought that he was dying and that his heart would beat no more. . . .

When he returned home, he quickly penned the following note: "I hold no one responsible for my death." Then he uttered a prayer, stretched himself out on his bed and drew a quilt over his head.

Till sunrise he lay awake waiting for his death and picturing to himself how his grave would be strewn with fresh flowers and how the birds would be singing above him. . . .

In the morning he was sitting on his bed and talking to Dashenka:

"Whoever lives a normal and temperate life, my dear friend, is immune from every harm, even poison. Take me, for example; I was already standing with both feet in the grave, I suffered, I died— and now, low and behold! I only feel a slight bitter taste in the mouth and my throat is a little burned; but as for my entire body, why, thank God. . . . And why? Because I lead a normal and decent life."

"Not at all! This only shows that my kerosene was worthless!" sighed Dashenka, thinking only of the price it cost. "This merely shows that the grocer, instead of giving me his best kerosene, gave me the stuff that costs only a cent and a half a quart! Good God, what robbers those people are! How they will take advantage of a poor, helpless woman! Thieves, murderers! May they live in the next world as peacefully as they let me live here! What blood-suckers they are! . . ."

And her tongue gathered more and more steam as it galloped on and on and on.

—Translated by Isaac Goldberg and
Henry T. Schnittkind

♈

ROBERT BURNS

(S C O T T I S H , 1 7 5 9 – 1 7 9 6)

From "Scotch Drink"

Gie him strong Drink *until he wink,*
That's sinking in despair;
An' liquor guid, to fire his bluid,
That's prest wi' grief an' care:

There let him bowse an' deep carouse,
Wi' bumpers flowing o'er,
Till he forgets his loves or debts,
An' minds his griefs no more.

Solomon's Proverbs,
Ch. 31ˢᵗ V. 6, 7.

Let other Poets raise a fracas
'Bout vines, an' wines, an' druken *Bacchus,*
An' crabbed names an' stories wrack us,   torment
           An' grate our lug,   vex; ear
I sing the juice *Scotch bear* can mak us,   barley
           In glass or jug.

O thou, my MUSE! guid, auld SCOTCH DRINK!
Whether thro' wimplin worms thou jink,   winding; frolic
Or, richly brown, ream owre the brink,   cream
           In glorious faem,   foam
Inspire me, till I *lisp* an' *wink,*
           To sing thy name!

Let husky Wheat the haughs adorn,   hollows
And Aits set up their awnie horn,   oats; bearded
An' Pease an' Beans, at een or morn,
           Perfume the plain,
Leeze me on thee *John Barleycorn,*   Blessings on thee
           Thou king o' grain!

On thee aft Scotland chows her cood,
In souple scones, the wale o' food!   best choice
Or tumbling in the boiling flood
           Wi' kail an' beef;   kale
But when thou pours thy strong *heart's blood,*
           There thou shines chief.

Food fills the wame, an' keeps us livin: belly
Tho' life's a gift no worth receivin,

When heavy-dragg'd wi' pine an' grievin;
But oil'd by thee,
The wheels o' life gae down-hill, scrievin,   *careering*
Wi' rattlin glee.

Y

JOHN CIARDI

(AMERICAN, 1916–1986)

Vodka

Vodka, I hope you will note is
upwind from all other essences.
Drink it all night and all day
and your aunt's minister could
not track you to perdition, not
even with his nose for it. Vodka
has no breath. Call it the dead-
man's drink. But praise it. As
long as he can stand, a vodka-
drinker is sober, and when he
falls down he is merely sleepy.
Like poetry, vodka informs anything with which it is diluted,
and like poetry, alas, it must be
diluted. Only a Russian can take
it straight, and only after long
conditioning, and just see what
seems to be coming of that!

Y

CHARLES DICKENS

(ENGLISH, 1812–1870)

From *Sketches by Boz,* Chapter 22

GIN-SHOPS

It is a remarkable circumstance, that different trades appear to partake of the disease to which elephants and dogs are especially liable, and to run stark, staring, raving mad, periodically. The great distinction between the animals and the trades is, that the former run mad with a certain degree of propriety—they are very regular in their irregularities. We know the period at which the emergency will arise, and provide against it accordingly. If an elephant run mad, we are all ready for him—kill or cure—pills or bullets, calomel in conserve of roses, or lead in a musket-barrel. If a dog happen to look unpleasantly warm in the summer months, and to trot about the shady side of the streets with a quarter of a yard of tongue hanging out of his mouth, a thick leather muzzle, which has been previously prepared in compliance with the thoughtful injunctions of the Legislature, is instantly clapped over his head, by way of making him cooler, and he either looks remarkably unhappy for the next six weeks, or becomes legally insane, and goes mad, as it were, by Act of Parliament. But these trades are as eccentric as comets; nay, worse, for no one can calculate on the recurrence of the strange appearances which betoken the disease. Moreover, the contagion is general, and the quickness with which it diffuses itself, almost incredible.

We will cite two or three cases in illustration of our meaning. Six or eight years ago, the epidemic began to display itself among the linen-drapers and haberdashers. The primary symptoms were an inordinate love of plate-glass, and a passion for gas-lights and gilding. The disease gradually progressed, and at last attained a fearful height. Quiet dusty old shops in different parts of town, were pulled down; spacious premises with stuccoed fronts and gold letters, were erected instead; floors were covered with Turkey carpets; roofs supported by massive pillars; doors knocked into windows, a dozen squares of glass into one; one shopman into a dozen; and there is no knowing what would have been done, if it had not been fortunately discovered, just in time, that the Commissioners of Bankruptcy were as competent to decide such cases as the Commissioners of Lunacy, and that a little confinement and gentle examination did wonders. The disease abated. It died away. A year or two of comparative tranquillity ensued. Suddenly it burst out again amongst the chemists; the symptoms were the same, with the addition of a strong desire to stick the royal arms over the shop-door, and a great rage for mahogany, varnish, and expensive floor-cloth. Then the hosiers were infected, and began to pull down their shop-fronts with frantic recklessness. The mania again died away, and the public began to congratulate themselves on its entire disappearance, when it burst forth with tenfold violence among the publicans, and keepers of "wine vaults." From that moment it has spread among them with unprecedented rapidity, exhibiting a concatenation of all the previous symptoms; onward it has rushed to every part of town, knocking down all the old public-houses, and depositing splendid mansions, stone balustrades, rosewood fittings, immense lamps, and illuminated clocks, at the corner of every street.

The extensive scale on which these places are established, and the ostentatious manner in which the business of even the smallest among them is divided into branches, is amusing. A handsome plate of ground glass in one door directs you "To the Counting-house;" another to the "Bottle Department;" a third to the "Wholesale Department;" a fourth to "The Wine Promenade;" and so forth, until we are in daily expectation of meeting with a "Brandy Bell," or a "Whiskey Entrance." Then, ingenuity is exhausted in devising attractive titles for the different descriptions of gin; and the dram-drinking portion of the community as they gaze upon the gigantic black and white announcements, which are only to be equalled in size by the figures beneath them, are left in a state of pleasing hesitation between "The Cream of the Valley," "The Out and Out," "The No Mistake," "The Good for Mixing," "The Real Knock-me-down," "The celebrated Butter Gin," "The regular Flare-up," and a dozen other, equally inviting and wholesome *liqueurs*. Although places of this description are to be met with in every second street, they are invariably numerous and splendid in precise proportion to the dirt and poverty of the surrounding neighbourhood. The gin-shops in and near Drury Lane, Holborn, St. Giles's, Covent Garden, and Clare Market, are the handsomest in London. There is more of filth and squalid misery near those great thoroughfares than in any part of this mighty city.

We will endeavour to sketch the bar of a large gin-shop, and its ordinary customers, for the edification of such of our readers as may not have had opportunities of observing such scenes; and on the chance of finding one well suited to our purpose, we will make for Drury Lane, through the narrow streets and dirty courts which divide it from Oxford Street, and that classical spot adjoining the brewery at the bottom of Tottenham Court Road, best known to the initiated as the "Rookery."

The filthy and miserable appearance of this part of London can hardly be imagined by those (and there are many such) who have not witnessed it. Wretched houses with broken windows patched with rags and paper: every room let out to a different family, and in many instances to two or even three—fruit and "sweet-stuff" manufacturers in the cellars, barbers and red-herring vendors in the front parlours, cobblers in the back; a bird-fancier in the first floor, three families on the second, starvation in the attics, Irishmen in the passage, a "musician" in the front kitchen, and a charwoman and five hungry children in the back one—filth everywhere—a gutter before the houses and a drain behind—clothes drying and slops emptying, from the windows; girls of fourteen or fifteen, with matted hair, walking about barefoot, and in white great-coats, almost their only covering; boys of all ages, in coats of all sizes and no coats at all; men and women, in every variety of scanty and dirty apparel, lounging, scolding, drinking, smoking, squabbling, fighting, and swearing.

You turn the corner. What a change! All is light and brilliancy. The hum of many voices issues from that splendid gin-shop which forms the commencement of the two streets opposite; and the gay building with the fantastically ornamented parapet, the illuminated clock, the plate-glass windows surrounded by stucco rosettes, and its profusion of gas-lights in richly-gilt burners, is perfectly dazzling when contrasted with the darkness and dirt we have just left. The interior is even gayer than the exterior. A bar of French-polished mahogany, elegantly carved, extends the whole width of the place; and there are two side-aisles of great casks, painted green and gold, enclosed within a light brass rail, and bearing such inscriptions as "Old Tom, 549;" "Young Tom, 360;" "Samson, 1421"—the figures agreeing, we presume, with "gallons," understand. Beyond the bar is a lofty and spacious saloon, full of the same enticing vessels, with a gallery running round it, equally well furnished. On the counter, in addition to the usual spirit apparatus, are two or three little baskets of cakes and biscuits, which are carefully secured at the top with wicker-work, to prevent their contents being unlawfully abstracted. Behind it are two showily-dressed damsels with large necklaces, dispensing the spirits and "compounds." They are assisted by the ostensible proprietor of the concern, a stout coarse fellow in a fur cap, put on very much on one side to give him a knowing air, and to display his sandy whiskers to the best advantage.

The two old washerwomen, who are seated on the little bench to the left of the bar, are rather overcome by the headdresses and haughty demeanour of the young ladies who officiate. They receive their half-quartern of gin and peppermint, with considerable deference, prefacing a request for "one of them soft biscuits," with a "Jist be good enough, ma'am." They are quite astonished at the impudent air of the young fellow in a brown coat and bright buttons, who, ushering in his two companions, and walking up to the bar in as careless a manner as if he had been used to green and gold ornaments all his life, winks at one of the young ladies with singular coolness, and calls for a "kervorten and a three-out-glass," just as if the place were his own. "Gin for you, sir?" says the young lady when she has drawn it: carefully looking every way but the right one, to show that the wink had no effect upon her. "For me, Mary, my dear," replies the gentleman in brown. "My name an't Mary as it happens," says the young girl, rather relaxing as she delivers the change. "Well, if it an't, it ought to be," responds the irresistible one: "all the Marys as ever *I* see, was handsome gals." Here the young lady, not precisely remembering how blushes are managed in such cases, abruptly ends the flirtation by addressing the female in the faded feathers who has just entered, and who, after stating explicitly, to prevent any subsequent misunderstanding, that "this gentleman pays," calls for "a glass of port wine and a bit of sugar."

Those two old men who came in "just to have a drain," finished their third quartern a few seconds ago; they have made themselves crying drunk; and the fat comfortable-looking elderly women, who had "a glass of rum-scrub" each, having chimed in with their complaints on the hardness of the times, one of the women has agreed to stand a glass round, jocularly observing that "grief never mended no broken bones, and as good people's wery scarce, what I says is, make the most on 'em, and that's all about it!" a sentiment which appears to afford unlimited satisfaction to those who have nothing to pay.

It is growing late, and the throng of men, women, and children, who have been constantly going in and out, dwindles down to two or three occasional stragglers—cold, wretched-looking creatures, in the last stage of emaciation and disease. The knot of Irish labourers at the lower end of the place, who have been alternately shaking hands with, and threatening the life of each other, for the last hour, become furious in their disputes, and finding it impossible to silence one man, who is particularly anxious to adjust the difference, they resort to the expedient of knocking him down and jumping on him afterwards. The man in the fur cap, and the potboy rush out; a scene of riot and confusion ensues; half the Irishmen get shut out, and the other half get shut in; the potboy is knocked among the tubs in no time; the landlord hits everybody, and everybody hits the landlord; the barmaids scream; the police come in; the rest is a confused mixture of arms, legs, staves, torn coats, shouting, and struggling. Some of the party are borne off to the station-house, and the remainder slink home to beat their wives for complaining, and kick the children for daring to be hungry.

We have sketched this subject very slightly, not only because our limits compel us to do so, but because, if it were pursued farther, it would be painful and repulsive. Well-disposed gentlemen, and charitable ladies, would alike turn with coldness and disgust from a description of the drunken besotted men, and wretched broken-down miserable women, who form no inconsiderable portion of the frequenters of these haunts; forgetting, in the pleasant consciousness of their own rectitude, the poverty of the one, and the temptation of the other. Gin-drinking is a great vice in England, but wretchedness and dirt are a greater; and until you improve the homes of the poor, or persuade a half-famished wretch not to seek relief in the temporary oblivion of his own misery, with the pittance which, divided among his family, would furnish a morsel of bread for each, gin-shops will increase in number and splendour. If Temperance Societies would suggest an antidote against hunger, filth, and foul air, or could establish dispensaries for the gratuitous distribution of bottles of Lethe-water,* gin-palaces would be numbered among the things that were.

Y

OGDEN NASH

(AMERICAN, 1902–1971)

A Drink with Something in It

There is something about a Martini,
A tingle remarkably pleasant;
A yellow, a mellow Martini;
I wish that I had one at present.
There is something about a Martini,
Ere the dining and dancing begin,
And to tell you the truth,

*From the mythical Greek river of forgetfulness that separates the kingdoms of life and death.

It is not the vermouth—
I think that perhaps it's the gin.

There is something about an old-fashioned
That kindles a cardiac glow;
It is soothing and soft and impassioned
As a lyric by Swinburne or Poe.
There is something about an old-fashioned
When dusk has enveloped the sky,
And it may be the ice,
Or the pineapple slice,
But I strongly suspect it's the rye.

There is something about a mint julep.
It is nectar imbibed in a dream,
As fresh as the bud of the tulip,
As cool as the bed of the stream.
There is something about a mint julep,
A fragrance beloved by the lucky.
And perhaps it's the tint
Of the frost and the mint,
But I think it was born in Kentucky.

There is something they put in a highball
That awakens the torpidest brain,
That kindles a spark in the eyeball,
Gliding singing through vein after vein.
There is something they put in a highball
Which you'll notice one day, if you watch;
And it may be the soda
But judged by the odor,
I rather believe it's the Scotch.

Then here's to the heartening wassail,
Wherever good fellows are found;
Be its master instead of its vassal,
And order the glasses around.
For there's something they put in the wassail

That prevents it from tasting like wicker;
Since it's not tapioca,
Or mustard, or mocha,
I'm forced to conclude it's the liquor.

♈

CHARLES FENNO HOFFMAN

(AMERICAN, 1806–1884)

The Origin of Mint Juleps

> And first behold this cordial Julep
>   here,
> That flames and dances in its crystal
>   bounds,
> With spirits of balm and fragrant syr-
>   ups mixed;
> Not that Nepenthes which the wife
>   of Thone*
> In Egypt gave to Jove-born Helena,
> Is of such power to stir up Joy as this,
> To life so friendly, or so cool to thirst.
>
>   MILTON—*Comus*.

'Tis said that the gods, on Olympus of old,
   (And who the bright legend profanes with a doubt?)
One night, 'mid their revels, by Bacchus were told
   That his last butt of nectar had somehow run out!

But, determined to send round the goblet once more,
   They sued to the fairer immortals for aid

---

* Polydamna, the wife of the Egyptian Thone, gave Nepenthes as a pain reliever to Helen
and Menelaus on their way home from Troy (*Odyssey, IV*).

In composing a draught, which, till drinking were o'er,
   Should cast every wine ever drank in the shade.

Grave Ceres herself blithely yielded her corn,
   And the spirit that lives in each amber hued grain,
And which first had its birth from the dews of the morn,
   Was taught to steal out in bright dew-drops again.

Pomona, whose choicest of fruits on the board
   Were scatter'd profusely in every one's reach,
When called on a tribute to cull from the hoard,
   Express'd the mild juice of the delicate peach.

The liquids were mingled, while Venus looked on,
   With glances so fraught with sweet magical power,
That the honey of Hybla,* e'en when they were gone,
   Has never been missed in the draught from that hour.

Flora then, from her bosom of fragrancy, shook,
   And with roseate fingers press'd down in the bowl,
All dripping and fresh as it came from the brook,
   The herb whose aroma should flavour the whole.

The draught was delicious, each god did exclaim,
   Though something yet wanting they all did bewail;
But juleps the drink of immortals became,
   When Jove himself added a handful of hail.

---

*An ancient town in Sicily famed for honey.

Ŷ

H. L. MENCKEN

(AMERICAN, 1880–1956)

From *The Vocabulary of the Drinking Chamber*

I have in my archives perhaps forty or fifty such etymologies for *cocktail*, but can only report sadly that nearly all of them are no more than baloney. The most plausible that I have encountered was launched upon humanity by Stanley Clisby Arthur, author of "Famous New Orleans Drinks and How to Mix Them," a classical work. It is to the effect that the *cocktail* was invented, along about 1800, by Antoine Amédée Peychaud, a refugee from Santo Domingo who operated a New Orleans pharmacy in the Rue Royale. This Peychaud was a Freemason, and his brethren in the craft took to dropping in at his drugstore after their lodge meetings. A hospitable fellow, he regaled them with a toddy made of French brandy, sugar, water, and a bitters of a secret formula, brought from Santo Domingo. Apparently running short of toddy glasses, he served this mixture in double-ended eggcups, called, in French *"coquetiers."* The true pronunciation of the word was something on the order of *"ko-kayt-yay,"* but his American friends soon mangled it to *"cock-tay"* and then to *"cocktail."* The composition of the bitters he used remained secret, and they are known as Peychaud's to this day. His brandy came from the Sazerac du Forge et Fils distillery at Limoges, and its name survives in the Sazerac cocktail, though this powerful drug is now usually made of rye whiskey, with the addition of Peychaud's bitters, absinthe, lemon peel, and sugar.

As I have said, this etymology has more plausibility than most, and I'd like to believe it, if only to ease my mind, but some obvious question marks follow it. First, why didn't Arthur give his authorities? It is hard to believe that he remembered back to 1800

himself, and if there were intervening chroniclers, then why didn't he name them? And if he got his facts from original documents— say, in the old Cabildo—then why didn't he supply titles, dates, and pages? A greater difficulty lies in the fact that the searchers for the "Dictionary of American English" unearthed a plain mention of the *cocktail* in the Hudson (N.Y.) *Balance* for May 13, 1806, in which it was defined as "a stimulating liquor composed of spirits of any kind, sugar, water and bitters." How did Peychaud's invention, if it *was* his invention, make its way from New Orleans to so remote a place as Hudson, New York, in so short a time, and how did it become generalized on the way? At the start of its journey it was a concoction of very precise composition—as much so as the Martini or Manhattan of today—but in a very few years it was popping up more than a thousand miles away, with an algebraic formula, $x + C_{12}H_{22}O_{11} + H_2O + y$, that can be developed, by substitution, into almost countless other formulas, all of them making authentic cocktails. Given *any* hard liquor, *any* diluent, and *any* addition of aromatic flavoring, and you have one instantly. What puzzles me is how this massive fact, so revolutionary in human history and so conducive to human happiness, jumped so quickly from New Orleans to the Hudson Valley. It seems much more likely that the *cocktail* was actually known and esteemed in the Albany region some time before Peychaud shook up his first Sazerac on the lower Mississippi. But, lacking precise proof to this effect, I am glad to give that Mousterian soda jerker full faith and credit, and to greet him with huzzas for his service to humanity.

Cocktails are now so numerous that no bartender, however talented, can remember how to make all of them, or even the half of them. In the "Savoy Cocktail Book," published in 1930, the number listed is nearly seven hundred, and in the "Bartender's Guide," by Trader Vic, published in 1947, it goes beyond sixteen hundred. No man short of a giant could try them all, and nine-tenths of them, I believe, would hardly be worth trying. The same sound instinct that prompts the more enlightened minority of mankind to come in out of a thunderstorm has also taught it to confine its day-in-and-day-out boozing to about a dozen standard varieties— the *Martini*, the *Manhattan*, the *Daiquiri*, the *Side Car*, the *Orange Blossom*, the *Alexander*, the *Bronx*, and a few others. The lexicog-

rapher John Russell Bartlett, in the fourth edition of his "Dictionary of Americanisms," 1877, also listed the *Jersey* and the *Japanese,*
but neither survives except in the bartenders' guides, which no
bartender ever reads. The "Dictionary of American English" traces
the *Manhattan* only to 1894, but that is absurd, for I saw a justice
of the Supreme Court of the United States drink one in a Washington barroom in 1886. The others that I have mentioned, save
the *Martini* and the *Bronx,* are not listed in any dictionary at hand,
though millions of them go down the esophagi of one-hundred-
per-cent Americans every week, and maybe every day. A corre-
spondent tells me that the *Daiquiri* was invented by American engineers marooned at Daiquiri, near Santiago de Cuba, in 1898;
they ran out of whiskey and gin but found a large supply of pale
Cuban rum, and got it down by mixing it with lime juice. The
origin of the *Martini* is quite unknown to science, though I have
heard the suggestion that its name comes from that of Martin Lu-
ther. The origin of *Bronx* ditto; all that is known is that it preceded the *Bronx Cheer.* The origin of the *Alexander* ditto, despite
some fancy theories. The *Old-fashioned* is supposed to be the
grandfather of them all, and it really may be, for its formula greatly
resembles that of the Hudson *Balance* cocktail of 1806, but the
fact remains that Bartlett did not mention it in 1877 and that it is
not to be found in any earlier reference works, not even the
Congressional Globe, predecessor of the *Congressional Record.*

Some time ago, the St. Louis *Post-Dispatch,* a high-toned paper,
dug up a local historian who testified that the *highball* was named
by Lilburn McNair, a grandson of the first governor of Missouri
and a shining light of St. Louis bar society in the nineties. It seems
to be true that the name was first heard about that time, for *highball* was never applied before to a mixture of rye or bourbon and
soda or tap water, and Scotch whiskey did not come upon the
American market until the early nineties. (My father, trying his
first shot of it in 1894, carried on in a violent manner, and died,
four years later, believing that it was made by quack saloonkeepers
in their cellars, of creosote and sweet spirits of nitre.) But there
are old-timers in Boston who say that the first *highball* was shoved
across the bar at the Parker House there, and the late Patrick Gavin
Duffy, an eminent bartender of New York, claimed in his "The

Official Mixer's Manual" that he borned it at the old Ashland House in 1895. Why the name? Most of the authorities say that it arose from the fact that the bartenders of the nineties called a glass a *ball* and that *highball* flowed naturally from the fact that what was formerly a whiskey-and-soda needed a taller glass than a straight whiskey. But all the bartenders above eighty that I am acquainted with say that *ball* was never used for a glass. Other authorities report that *highball* was lifted from the railroad men, who use the term for go ahead. But this sounds pretty thin, for if the railroad men of that era had ever detected a bartender putting water (and especially soda water) into whiskey, they would have butchered him on the spot. Thus the matter stands. I pant for light, but there is no light.

The history of *Mickey Finn* is equally murky. Herbert Asbury says, in his "Gem of the Prairie," a history of the rise of culture in Chicago, that the name was borrowed from that of a Chicago saloonkeeper who had been a lush-roller in his early days and operated a college for pickpockets in connection with his saloon. The patrons of the place were a somewhat mischievous lot, and not infrequently Finn had to go to the aid of his bouncer. They used the side arms in vogue at the time—to wit, bung starters, shillelaghs, joints of gas pipe, and lengths of garden hose filled with BB shot—but the work was laborious, and Finn longed for something sneakier and slicker. One day, a colored swami named Hall offered to mix him a dose that would knock out the friskiest patient in a few minutes. The formula turned out to be half an ounce of chloral hydrate in a double slug of pseudo-whiskey. It worked so well that many of those to whom it was given landed in the morgue, and Finn was so pleased with it that he gave it his name. I have a very high opinion of Asbury's lexicographical and sociological parts, but I am still waiting to hear him explain how *Mickey Finn* became transferred from a dose comparable to an atomic bomb to a drink of bathtub gin with a drop or two of croton oil in it—a mixture that certainly got rid of the customer but did him no more permanent harm than a draught of Glauber's salts.* Also, I am waiting to hear from him why a Chicago saloon-keeper had to wait

---

*The cathartic made from seeds of the Asian *Croton tiglium* tree is compared with a common laxative, crystalline decahydrate of sodium sulfate.

for a colored necromancer to tell him about *knockout drops,* which had been familiar in American criminal circles since the first Grant administration. My own great-uncle, Julius by name, got a massive shot of them in Wheeling, West Virginia, in 1870, and was never the same man afterward.

These few examples reveal the pitfalls, booby traps, and other difficulties that strew the path of anyone seriously interested in the origin and history of booze terms. The dictionaries, always prissy, avoid most of them as they avoid the immemorial four-letter words. You will not find *Mickey Finn* in the great Webster of 1934, or in the "Dictionary of American English," or in the "Supplement to the New English Dictionary." It appears, to be sure, in some of the newer and smaller dictionaries, but almost always with the equivocal definition of "a drugged drink." So far as I have been able to discover, only "Words: The New Dictionary," brought out in 1947, says that its essential medicament is a cathartic, not a narcotic. Even Berrey and Van den Bark, in their invaluable "American Thesaurus of Slang," 1942, are content to list it under the rubric of *strong liquor,* along with *forty-rod, pop-skull,* and *third rail,* though I should add that they note that tramps and criminals now use it to designate any laxative victual. *Highball* is listed in nearly all the dictionaries published since 1930, but not one of them attempts its etymology. Nor does any of them try to unravel the mystery of *cocktail.*

♥

CHERYL PALLANT

(A M E R I C A N ,   1 9 6 0 -)

A Neighborhood Bar

John walked into the smoke-filled bar with soft lights, a rest from the harshness otherwise surrounding his life. He ordered himself a drink and surveyed the patrons. They were already tipsy, not walking as straight as when they first entered the bar, or too straight, parading around the room as if imitating a two-by-four.

Glass in hand, he steered himself through the crowd, stopping for none of the prattle endemic to the room. Bitterly and sometimes self-righteously, his attitude reproached them, thereby elevating himself. They were too settled and he liked to be stirred, at least occasionally. His ultimate conclusion—they weren't worth his time. Merely, he sauntered past their stylish bodies with the ease of movement expected of a man whose every pint of blood is filled with several ounces of molten lead.

He wanted to remain alone and sought out no one's company.

Except one person. Only she, Margaretta (so he thought her name to be), would he speak to. Only she, a woman of unique dress and, he imagined, a salty personality, would cause him to break the vow he'd made in response to his brother eight years ago. Because of an argument concerning a woman, John had promised not to engage in conversation with others except when necessary. He rationalized; this was necessary.

For the past month, he had seen Margaretta here, always alone. No one at her side when she walked in, she left the same way, no one clinging to her or escorting her elegant body out the door. Many men tried to change her circumstance. None succeeded despite varied attempts.

John hoped to see her tonight. He would break his vow of silence. He would enter the mass of longing men and pour out his desires, hoping she could contain them. He was different from other men; in this, he drew confidence. She would recognize his unique value, that he was a man of taste whose sobriety could easily lead to unabashed delirium.

The evening lengthened. He kept his eye keen on the entrance. Still, she had not arrived. Would she?

He finished his third drink.

In waiting, opportunities presented themselves and moved along. People tried to converse with him: "And what do you do?" "Is that a Calvin Klein shirt?" "You wouldn't happen to know the weather for tomorrow by chance, would you?" He answered curtly and the questioners, turned off, turned away.

Tired of being on his feet, he seated himself on a stool at the bar. He avoided eye contact. His impatient, single-minded voice overpowered the backdrop of chattering voices and pulsing music. Occasionally, the sound of the saxophone slithered into his consciousness but quickly assimilated into the blend surrounding him.

He ordered a fourth drink. With his back to the bar, elbows propped on its wooden edge, he stared out. Would she ever come? He so wanted her to.

She did.

Unnerved by the sighting, he nearly spilled his drink on his pants. What would he say to her?

She made her way to the bar and sat several stools away from his. Unrehearsed, he strolled toward her. He wanted to meet her immediately, to absorb her in conversation. Already, he'd been waiting too long.

As suavely as possible, one hand elegantly wrapped around his drink, the other casually looped from his pocket, he approached her. He tossed his head back, pushing hair away from his forehead.

Then he fell.

His drink crashed to the floor along with his stunned body. He was afraid to look up, to see whose eyes looked upon him. He eyed the floor in misery with a face that once hinted at glory. What had happened? What made him descend so abruptly to this level, beside everyone's feet? He was enraged. He considered banging the floor with his fist, a moment of tantrum, but resisted.

Then he noticed the culprit, a rusty nail sticking up about an inch from a floor board. It had gashed a hole in his pants and his leg. Revenging himself on the piece of metal with the ferocity of a wild animal, he tore it out from its once secure hold. He waved it above his head, planning to use it as evidence in suing the bar.

While he stood, faces gawked at him, some with concern, some with glee, he the unannounced spectacle who had taken nearly the entire audience away from the band. He cared only about seeing one face. Hers. Had she seen him? She must have.

"Where's the manager?! I demand to see the manager!" From the cloud of his huff appeared an obedient servant, a waiter, who carried him off as if by flight into a backroom.

Several men and women lining the room on burgundy leather couches and chairs watched him quietly as he walked in. Then he saw her, Margaretta. She rocked in a chair behind a large black desk.

"You!" Astonishment escaped him.

"Yes. You wanted to see me?"

"Who are they?" Their presence unnerved his already frazzled responses.

"Friends." They had the look of a mafia.

"I suppose you want to sue?" She leaned back confidently, indifferently. "I can pay for your pants. What more do you want?"

"Well . . ."

"Just as I expected. How about a deal?" She paused. "I'd like to make you one of us."

"What!"

"I know all about you, John Collins, and I also know about your brother, Tom."

"Tom?"

She pushed locks of her hair away from her shoulder. "Yes, Tom. And I know about his involvement with Rob Roy, Harvey Wallbanger, 'King' Alfonce and 'Bloody' Mary."

"What!" At that moment, John, recognizing the muddle he was in, knew he had no choice in the matter. He would have to do exactly what she requested. Exactly. He loathed hearing himself say the word. It chilled him. He'd been reduced, cut down, his life changed. How quickly romance moves into different phases.

"What do you want from me?"

"To work with your brother and his friends. As a mixer."

Grudgingly, John met up with Tom and was forced to work at a network of bars to ensure patrons remain blissfully amnesiac, just this side of dipsomania.

Y

BERTOLT BRECHT

(GERMAN, 1898 – 1956)

Exemplary Conversion of a Grog-Seller

1

Bottles, glasses, on the bar behind him
Heavy-lidded, lips of violet
Dreary eyes in his perspiring features
Sits a grog-seller pallid and fat.
His greasy fingers count the money
Pushing coins into a heap:
Then in an oily pool of gin he
Drops his head and falls asleep.

2

And his body heaves, he writhes there groaning
Cold sweat smears his forehead like slime
While in his spongelike brain there enter
Nightmare figures as in pantomine.
And he dreams: he is in heaven
And must go where God's enthroned
He because the thought unmans him
Drinks till he's completely stoned.

3

Seven angels form a ring around him
And he starts to stagger on his feet
But the publican is led reluctant
Speechless to God's judgement seat.
He can't raise his heavy eyelids
Of God's white light he stands in dread

Feels his tongue is blue and stuck there
Weighing in his mouth like lead.

4

And he looks round in search of rescue
And he sees in green and seaweed light:
Fourteen orphan children floating slowly
Downstream, faces fading ashen white.
And he says: it's only seven.
Being drunk I see them twice.
Cannot say it, since his tongue is
Firmly held as in a vice.

5

And he looks round in search of rescue
Sees men who play poker all day long
And he bawls: I am your friend the grog-seller!
They go on bawling their drunken song.
And they bawl that their salvation
Lies in whisky or in gin.
And he sees their pale green blotches:
Putrefaction has set in.

6

And he looks round in search of rescue
And he finds: I'm not wearing what I think.
I'm in underpants in heaven! Hears their question:
Did you sell off all your clothes for drink?
And he says: I did have clothes once.
And they say: Don't you feel shame?
And he knows: Many have stood here
Whom I have stripped of all but name.

7

And he no longer looks for rescue
And he kneels so quickly his knees crack
And he feels the sword where neck meets body
And the shirt sweat-damp against his back:
And he feels the scorn in heaven

And deep within he is aware:
Because I am a grog-seller
God has banished me from there.

8

Then he wakes: though with eyes still bleary
Heavy-lidded, lips of violet.
But he tells himself: No longer will I
Be a grog-seller pallid and fat.
Rather will I seek out orphan children
Drunks, the old, the chronic ill
They alone shall henceforth get this
Filthy lucre from the till.

—*Translated by H. B. Mallalieu*

Y

LÁL-BEHÁRI DAY

(I N D I A N ,   1 8 2 4 – 1 8 9 4)

From *Bengal Peasant Life*

THE VILLAGE GROG-SHOP

Pass where we may, through city or
  through town,
Village, or hamlet, of this merry land,
Though lean and beggared, every
  twentieth pace
Conducts the unguarded nose to such
  a whiff
Of stale debauch, forth issuing from
  the styes
That law has licensed, as makes Temperance reel.

—*William Cowper*

Some days after Govinda's big hut had been reduced to ashes, Babu Jaya Chánd Raya Chaudhuri was sitting in his Cutcherry, or rather half sitting, half lying down, his elbows resting on a huge bolster, and smoking, by means of one of those monstrous pipes called snakes, "which extended long and large, lay floating many a rood"; when Jaya Chánd was sitting in this interesting position, surrounded by the pomp and circumstance of landlordism, by his divána, his gomasṭa, and mohurirs; Bhima Sardár, the captain of his clubmen, stood in front of the room, and made a profound bow. Jaya Chánd raised his head a little, took off the snake from his mouth, and said, "Well, Bhima Sardár, what's the news?"

Bhim. "Maháráj! everything is right. In Maháraja's dominions nothing can go wrong."

Jaya. "You managed the thing rather neatly, the other night. But you were almost caught, and if the fellow had once seized you, it would have been all over with you; for the rascal is not only as black but as strong as Yama."*

Bhim. "That fellow catch me! Maháráj, by your blessing I have strength enough to fight alone and unaided with half-a-dozen fellows like him."

Jaya. "Ah, well, I don't know about that; you are exaggerating your strength. That fellow must be stronger than you. If you are Bhim, he is Arjjuna. However, you did your work successfully. You deserve some *bakshish* (reward)."

Bhim. "Maháráj! everything I have is your lordship's. In your lordship I live and eat and drink. What more could I expect?"

Jaya Chánd ordered the treasurer to give to Bhim Sardár two rupees. As the treasurer threw the rupees on the floor, the zamindár said to the Sardár, "Take that *táká,* and enjoy yourself with your companions." The Sardár made another profound bow, repeating the words, "Ráma, Ráma! Maháráj," and went away.

Leaving Jaya Chánd and his ministers to their confabulations, with the reader's permission, we will accompany the Sardár and see how he and his friends enjoy themselves. He, along with about a dozen of his companions, went to a grocer's shop and bought some quantities of *muḍi, muḍki, bátás, phuṭ-kalái,* and *páṭáli,*† and

*Identified as the first man in Hindu mythology. Because he was the first to die, he became Lord of the Dead.
†Typical snacks eaten with drinks.

went towards the tank Krishna Ságara, below the high embankment of which lay the only grog-shop in the village.

I suppose there were grog-shops in the country before its occupation by the British, but there is no doubt that the increase of drunkenness in the land is chiefly owing to the operation of the Ábkári system of the Government. That three thousand years ago the Aryan settlers in India were, as may be inferred from the Rigveda, hard drinkers and staunch beef-eaters, is probably true, but it is equally true that the inhabitants of India have been for centuries the soberest people in the world; and it is sad to think that amongst this remarkably temperate people drunkenness should be introduced and extended by the foremost Christian nation in the world. It is all very well to say that the excise on spirits, which it is the office of the Ábkári Department to collect, has a tendency to repress and not to increase the sale of spirits, as it makes them dearer than before; but anyone who is acquainted with the working of that department must be aware that the practice belies the theory. The aim of the Ábkári, like that of every other branch of the fiscal service, is to increase the public revenue; but the revenue from spirits cannot be increased without an increase in their sale. It is therefore the aim of all Ábkári officers to establish as many grog-shops as they can in the country. To this State interference we owe it that there are at this moment, in the year of grace 1871, more grog-shops in the country than it ever had since it was upheaved from the universal ocean. Now every village almost has a grog-shop, and the larger villages more than one; though things were not in so bad a state in those days when the scene of this story is laid as at present.

But to proceed. Bhim Sardár and his dozen friends went to the grog-shop, which was a small mud hut thatched with straw. They sat on the ground and ordered some *kalsis* of an intoxicating liquid called *dheno,* prepared from *dhán,* that is, paddy.* In those days no European wines, spirits, or liquors could be had in the village grog-shop. Only two sorts were sold, one prepared from treacle, and the other from paddy. The spirit distilled from treacle was rather dear for the poorest classes, who alone drank, as it was sold about eight annas or a shilling a bottle; whereas *dheno,* or, as it

*Unmilled, rough rice.

used to be jocosely called, *dhányesvari* (that is, the goddess of paddy), which is simple fermented paddy and water, a gallon could be had for a trifle. A large *haṇḍi* of *dhányesvari*, which is as much as any human stomach can contain, was sold for one *payasá*, which is a little less than a halfpenny. The thirteen clubmen, including Bhima Sardár himself, sat in a row on the floor on their feet, without their bodies couching the ground, lifting up their heads and keeping their mouths open, as if they were going to catch the rain. The *sunḍi* (wine-seller), with a *kalsi* of *dhányesvari* in his hand, approached them, and into the open mouth of the first person in the row he poured as much liquid as he wished to drink. In this manner the whole company, thirteen in number, were served. After all had had a drink, they opened their store of provisions, which consisted of *muḍi* and *muḍki, bátása, phut-kalái,* and *pátáli,* and fell upon them with the utmost fury. The *muḍi* and *muḍki* were in such large quantities that when put on a piece of cloth on the floor they made a hillock. But in a short time the whole was demolished. As they ate the *phut-kalái* they became very jolly, and cracked many a joke; some of them rolled on the mud floor, and thus gave unmistakeable signs of intoxication. After the provisions had been all demolished, they again called for *dhányesvari,* and the *sunḍi* again served them in the manner I have already described, pouring into their upraised throats as much of the liquid as they wanted. They now prepared their hookahs and smoked tobacco,

> Sooty retainer to the vine,
> Bacchus' black servant, negro fine.

With hookah in hand, some half-tipsy, and some entirely so, they left the grog-shop, and passed through the village, dancing and wild with merriment. I have heard it said by some of the inhabitants of Kánchanpur, that of an evening respectable Bráhmans of the village sometimes visited the grog-shop. I believe the statement is true, but still there is no doubt that the grog-shop was visited mainly by the lowest class of inhabitants, a class lower than that to which our hero belonged.

♟

RUSSIAN FABLE

The Drunkard's Vow

A toper made a solemn vow he never more would touch,
Or punch, or grog, or spirits mixed, or any compounds such.
Yet though to make 't was easy, to keep so strict a vow,
To prove an easy matter was not likely, you'll allow.
Soon after was our tippler seen reeling 'long the street.
"How now!" a neighbour cried, "why, you scarce can keep your
    feet.
I thought you had forsworn for ever punch and grog?"
"And so I have, nor do I now touch either, you dull dog;
But I keep my vow unbroken by drinking spirits *neat*."

♟

ROBERT GRAVES

(ENGLISH, 1895–1985)

Ouzo Unclouded

Here is ouzo (she said) to try you:
Better not drowned in water,
Better not chilled with ice,
Not sipped at thoughtfully,
Nor toped in secret.
Drink it down (she said) unclouded
At a blow, this tall glass full,
But keep your eyes on mine
Like a true Arcadian acorn-eater.

Y

J. K. HUYSMANS

(FRENCH, 1848–1907)

From *Against The Grain*

He returned to the dining room and made the melancholy comparison of himself in close quarters to voyagers struck by sea sickness in their cabins; he steered himself, stumbling toward the cupboard, scrutinized the mouth organ but never opened it, and seized from the highest shelf a bottle of Benedictine which he had been saving because its shape suggested to him ideas at once sweetly seductive and vaguely mystical.

But for the moment he remained indifferent, observing with a dull eye the somber green trappist bottle which at other times invoked medieval monastaries by its old monkish paunch, its head and neck dressed in a parchment cowl, its red wax seal quartered with three silver mitres on an azure field, and tied at the throat by lead bonds like a papal bull with its label written in sonorous Latin on yellow paper as if faded with time: *liquor Monachorum Benedictinorum Abbatiae Fiscanensis*.

Under this positively monastic habit signed with a cross and ecclesiastical initials D.O.M.,* wrapped in its parchment and ligatures just like an authentic charter, there slept a saffron-colored liqueur of exquisite delicacy. It exuded the quintessential aroma of angelica and hyssop mingled with seaweed that had been sweetened to mask the iodine and bromine; and it stimulated the palate with a spiritual passion that was concealed beneath a daintiness altogether virginal, like a novice, flattering the nostrils with a hint of corruption engulfed in a caress that was at the same time both childlike and passionately devout.

This hypocrisy, which resulted from the extraordinary discord between the container and the contained, between the liturgical

---

Deo Optimo Maximo (To God, Most Good, Most Great).

contours of the bottle and its soul, altogether feminine, altogether modern, had provoked his reveries before now. Long ago he had also dreamed, sitting before this bottle, of the monks themselves who sold the liqueur, the Benedictines at Frécamp Abbey, who, belonging to the congregation of Saint Maur, famous for its historical research, were subject to the rule of St. Benedict, without following either the observances of the white monks of Cîteaux or the black monks of Cluny.* Invincible, they appeared before his imagination, as if out of the Middle Ages, cultivating their herbs, heating their retorts, and, with incontestible authority, distilling in their alembics their sovereign panaceas.

He took a sip of the liqueur and, for a few moments, felt comforted. . . .

—*Translated by Joan Digby*

ALPHONSE DAUDET

(FRENCH, 1840–1897)

The Elixir

"Drink this, neighbour, and tell me what you think of it."

And drop by drop, with the scrupulous care of a lapidary counting pearls, the curé of Graveson poured me out two fingers of a green, golden, warm, sparkling, exquisite liqueur. It brought a flood of sunshine into my stomach.

"It is Père Gaucher's elixir, the joy and the health of our Provence," said the good man, with a triumphant air; "it is made at the convent of Prémontrés, a couple of leagues from your mill. Is it not worth all the Chartreuse in the world? And if you knew how amusing the story of this elixir is! Just listen!"

*Cistercian monastery founded on the Côte d'Or in 1098 by Saint Robert; Benedictine order founded at Mâcon-sur-Saône in 912.

Then, quite simply, and without seeing the joke of it, the abbé began in the dining-room of the *presbytère,* so quiet and calm, with its Way of the Cross in little pictures, and its pretty white curtains starched like surplices, a somewhat sceptical and irreverent story, of the fashion of a tale by Erasmus or D'Assoucy.

Twenty years ago the Prémontrés—*or rather the White Fathers, as our Provençaux call them—had become wretchedly poor. If you had seen their house in those days, you would have pitied them.

The great wall and the Pacôme Tower were crumbling away. All about the grass-grown cloister the colonnades were falling, and the stone saints toppling over in their niches. There was not a whole window, or a door which would shut. The Rhône wind blew in the closes and in the chapels, extinguishing the candles, breaking the lead of the casements, and spilling the holy water from the basins. But saddest of all was the convent belfry, as silent as an empty dove-cote; and the fathers, for lack of money to buy a bell, were obliged to ring matins on rattles made of almond wood!

Poor White Fathers! I can still see them at the Corpus-Christi procession, marching sadly in their patched gowns, pale, thin, fed on a diet of lemons and watermelons; and behind them the abbot, who walked with hanging head, ashamed to show his ungilded crosier and his moth-eaten mitre of white cloth. The ladies of the sisterhood wept with pity in the ranks, and the fat banner-bearer sneered in their midst under his breath as he pointed at the poor monks:

"Starlings go thin when they go in flocks."

The fact is that the unfortunate White Fathers had themselves reached the point of questioning whether it would not be better for them to take their flight into the world, each to seek his food in his own direction.

Well, one day when they were debating this grave question in the chapter, word was brought to the prior that Frère Gaucher asked to be heard by the council. You must know, for your better

*Monks of the Chanoine order founded by St. Norbert in 1120.

comprehension, that this Frère Gaucher was herdsman of the convent; that is to say, he spent his days wandering from arcade to arcade in the cloister, driving before him two lean cows which sought for grass in the cracks of the pavement. The poor herdsman, who had been cared for till the age of twelve by a crazy old woman of Baux* called Aunt Bégon, and who since then had been taken in by the monks, had never been able to learn anything but to drive his cattle and to say his Pater Noster; and even that he said in Provençal, for his brain was impenetrable, and his wit like a leaden dagger. For the rest he was a fervent Christian, though somewhat visionary, comfortable in his hair shirt, flagellating himself with hearty sincerity, and with such arms!

When they saw him enter the chapter-house, simple and awkward, bowing to the company with a scrape of the foot, the prior, canons, and treasurer all began to laugh. It was the effect he always produced whenever he went anywhere, by his good face with its grayish goat-like beard and his somewhat wild eyes, and therefore Frère Gaucher was not disturbed by it.

"Reverend Fathers," said he, good-humouredly, twisting his olive-wood rosary—"they are right when they say that empty hogsheads sing the loudest. By digging in my poor head, which is already so hollow, I think that I have discovered the means of getting us out of our difficulties.

"This is how. You remember Aunt Bégon, that good woman who took care of me when I was little? (May God keep her soul, the old sinner! she sang terrible songs when she had been drinking!) I must tell you then, Reverend Fathers, that Aunt Bégon knew as much about the herbs of our mountain as—yes, more than an old Corsican blackbird. Among other things she had composed toward the end of her life and incomparable elixir by mixing five or six kinds of simples which we used to go together and gather on the Alpilles. That is many years ago; but I think that with the help of Saint Augustine and the permission of our father the abbot, I might, if I tried hard, recover the composition of this mysterious elixir. Then we would have nothing to do but to put it into bottles and to sell it somewhat dear, which would allow

*Medieval town in the Bouches-du-Rhône Département.

the community to grow rich quietly, like our brothers of La Trappe and the Grande—"*

He had not time to finish. The prior had risen and fallen upon his neck. The canons clasped his hands. The treasurer, more deeply moved than any of the others, kissed with respect the tattered border of his robe. Then each one returned to his stall to deliberate, and on the spot the chapter decided that they would intrust the cows to Frère Thrasybule, in order that Frère Gaucher might give himself up wholly to the confection of his elixir.

How did the good brother succeed in recovering Aunt Bégon's receipt? At the cost of what efforts, what vigils? History does not say. Only, what is certain is that at the end of six months the elixir of the White Fathers was already very popular. In the whole country, in the whole neighbourhood of Arles, there was not a *mas*† nor a grange, which had not, in the depths of its cupboard, between the bottles of mulled wine and the jars of picholine olives, a little brown earthenware flagon, sealed with the arms of Provence, with a monk in ecstasy on a silver label. Thanks to the popularity of its elixir, the house of the Prémontrés grew rich very rapidly. They rebuilt the Pacôme Tower. The prior had a new mitre, the church pretty stained-glass windows; and within the fine lace of the belfry, a whole covey of bells, big and little, alighted one fine Easter morning, pealing merrily.

As for Frère Gaucher, that poor lay brother whose clownishness had so amused the chapter, he was no more heard of in the convent. Henceforth they knew only the Reverend Père Gaucher, the man of intellect and of great learning, who lived completely isolated from the many and trifling occupations of the cloister, and shut himself up all day in his distillery, while thirty monks scoured the mountain in search of fragrant herbs for him. This distillery, into which no one, not even the prior, had the right to enter, was an old, abandoned chapel, at the very end of the canon's garden. The simplicity of the good fathers had made of it something mysterious and terrible; and if by chance a bold and curious young monk, clinging to the climbing vines, reached the rose-window

* Other local monasteries; the first is a Trappist order.
† Farmhouse.

over the door, he would slip down hastily, frightened by the sight of Père Gaucher, with his wizard's beard, leaning over his furnaces, hydrometer in hand, and surrounded by retorts of rose-coloured sandstone, gigantic alembics, crystal worms, and a whole weird apparatus which shone as if bewitched in the red light of the windows.

At nightfall, when the last angelus rang, the door of this abode of mystery would be opened cautiously, and the reverend father would betake himself to the church for the evening service. What a reception they gave him whenever he walked through the monastery! The brothers would form a lane for him as he passed, and whisper—

"Hush! he has the secret."

The treasurer would follow him and speak to him in an undertone. Amidst all this adulation the father would walk on, mopping his brow, with his broad three-cornered hat on the back of his head like an aureole, looking about him complacently at the large courts planted with orange-trees, the blue roofs on which turned new weather-cocks, and, in the dazzlingly white cloister, between the elegant, florid columns, the canons clad in new robes, filing two and two with tranquil mien.

"It is to me that they owe it all!" the reverend father would tell himself; and each time this thought would bring a rush of pride.

The poor man was well punished for it. You will see how.

Just fancy that one evening, during service, he arrived at the church in a state of extraordinary excitement—red, out of breath, with his cowl awry, and so agitated that in taking holy-water he wet his sleeve to the elbow. They thought at first that it was embarrassment at arriving late; but when they saw him make low obeisances to the organ and the pulpit instead of to the high altar, cross the church like a whirlwind, and wander about the choir for five minutes in search of his stall, and then, when seated, bow right and left with a maudlin smile, a murmur of astonishment ran through the three naves. They whispered from breviary to breviary—

"What is the matter with our Père Gaucher? What is the matter with our Père Gaucher?"

Twice did the prior impatiently beat upon the pavement with

his crosier to command silence. Above, in the choir, the Psalms still went on, but the responses lacked enthusiasm.

Suddenly, in the very midst of the *Ave Verum,** our Père Gaucher suddenly sits back in his stall, and sings in a mighty voice—

> *Dans Paris il y a un Père Blanc,*
> *Patatin, patatan, tarabin, taraban.*†

There is general consternation. Every one rises. They cry— "Take him out! He is possessed!"

The canons cross themselves. The abbot's crosier keeps up a terrible clatter. But Père Gaucher neither sees nor hears anything; and two lusty monks are obliged to drag him out by the small choir door, struggling like one exorcised, and continuing his patatins and his tarabans louder than ever.

The next morning at daybreak the unhappy father was on his knees in the prior's oratory, saying his *mea culpa* with a torrent of tears.

"It is the elixir, Reverend Father, it is the elixir which took me by surprise," said he, beating his breast; and seeing him so wretched and repentant, the good prior was quite moved himself.

"Come, come, Père Gaucher, calm yourself; this will all dry up like dew in the sunshine. After all, the scandal was not so serious as you think. It is true that the song was rather—hum!—hum! Well we must hope that the novices did not hear it. And now tell me how it happened. It was in trying the elixir, was it not? Your hand was a little too heavy. Yes, yes, I understand. You are like Brother Schwartz, who invented gunpowder—the victim of your own invention. Tell me, my good friend, is it absolutely necessary for you to try this terrible elixir upon yourself?"

"Unfortunately, yes, Monseigneur, the test gives me the strength and degree of the alcohol; but for the finish, the velvet, I can trust only to my tongue."

"Ah! very well. But one thing more; when you taste the elixir in this way, by necessity, does it seem to you good? Do you enjoy it?"

---

*Call to duties.
†In Paris there is a White Father, and so on and so forth. . . .

"Alas! yes," said the unfortunate father, growing very red. "For the last two evenings it has seemed to have such a flavour, such an aroma! It is certainly the devil who has played me this trick; and therefore I am firmly resolved henceforth to use only the test. So much the worse if the liqueur is not fine enough, if it does not pearl—"

"Oh, Heaven forbid!" interrupted the prior hastily. "We must not risk displeasing our customers. All that you have to do, now that you are warned, is to be on your guard. Come, how much do you need to make sure? Fifteen or twenty drops, eh? Let us say twenty drops. The devil will be very sharp if he catches you with twenty drops. Besides, in order to prevent accidents, I will excuse you from coming to chapel hereafter. You can read the evening service in the distillery. And now, go in peace, Reverend Father, and above all, count your drops."

Alas! it was in vain that the poor father counted his drops; the devil had hold of him and never let go.

They were singular services that the distillery witnessed!

During the day all went well. The father was calm enough; he prepared his retorts and alembics, carefully selected his herbs—all the herbs of Provence, fine, gray, serrated, saturated with perfume and sunshine. But in the evenings, when the simples were infused, and the elixir was cooling in great pans of red copper, then the poor man's martyrdom would begin.

"Seventeen—eighteen—nineteen—twenty!"

The drops would fall from the graduator into the silver goblet. The father would swallow these twenty at a gulp, almost without pleasure. It was only for the twenty-first that he longed. Oh, that one-and-twentieth drop! Then, to escape from the temptation, he would go and kneel at the other end of the laboratory and plunge into his prayers. But from the still warm liqueur there would rise a little mist laden with aromatic odours which came and played about him, and in spite of himself drew him back to the pans. The liqueur was of a fine golden green. Bending over it, with open nostrils, the father would stir it gently, and in the little flashes which shone against the emerald background he seemed to see Aunt Bégon's eyes, laughing and sparkling as they looked at him.

"Come, one drop more!"

And, from drop to drop, the unfortunate man would end by having his goblet filled to the brim. Then, at the end of his strength, he would sink into a large easy-chair and with half-closed eyes lazily sip his sin, murmuring to himself with delicious remorse,

"Ah! I am damning myself, I am damning myself!"

The most terrible part of it was that in the depths of this diabolical elixir he would find, by some strange witchcraft, all Aunt Bégon's dreadful songs: *Ce sont trois petites commères qui parlent de faire un banquet;* or *Bergerette de Maître André s'en va-t-au bois seuletie;* * and invariably the famous one about the White Fathers—*Patatin, patatan.*

Fancy his embarrassment the next morning when those who occupied the neighbouring cells would say to him—

"Ah, ha! Père Gaucher, so you had cicadæ in your head last night when you went to sleep?"

Then there would be tears, despair, fasting, the hair shirt, and the scourge. But nothing could prevail against the demon of the elixir; and every evening, at the same hour, the poor father would be again possessed.

During this time orders were fairly showered upon the abbey like a benediction. They came from Nîmes, from Aix, from Avignon, from Marseilles. Day by day the convent took on more the air of a manufactory. There were packing brothers and labelling brothers, brothers for writing and brothers for carting. The service of God may have lost the tolling of a bell here and there; but the poor of the neighbourhood lost nothing by it, I warrant you.

Well, one fine Sunday morning, while the treasurer was reading to the assembled chapter his inventory for the close of the year, and the good canons were listening with glittering eyes and smiling lips, Père Gaucher suddenly rushed into the midst of the conference, crying—

"That ends it! I will make no more. Give me back my cows—"

"What is the matter, Père Gaucher?" asked the prior, who suspected what was in the wind.

"What is the matter, Reverend Father? The matter is that I am

---

*The one is about three little gossips who plan to make a presumably unholy feast; the other is about Master André's young shepherdess who goes to the woods alone.

preparing for myself a nice eternity of flames and pitchforks. The matter is that I am drinking—that I am drinking like a fish."

"But I told you to count your drops."

"Ah, yes, count my drops! It is by goblets that I should have to count now. Yes, Reverend Father, I have reached that point; three flasks an evening. You can understand that this cannot go on. Therefore let who you like make your elixir. May the fire of God burn me if I will have anything more to do with it!"

As you can fancy, the chapter was not smiling now.

"But you will ruin us, wretched man!" cried the treasurer, brandishing his ledger.

"Do you prefer to see me damned?"

Then the prior rose.

"Reverend Fathers," said he, stretching forth his white hand on which glistened the pastoral ring, "there is a way of arranging everything. It is in the evening, is it not, my dear son, that the demon tempts you?"

"Yes, regularly every evening. Consequently, now when I see evening approach, I sweat, saving your presence, from head to foot, like Capitou's ass when he saw the saddle coming."

"Well, take heart! Every evening hereafter, at service, we will recite for you the orison of Saint Augustine, to which plenary indulgence is attached. With that, no matter what happens, you are safe. It is absolution during the sin."

"Oh! very well then; thank you, Reverend Father!"

And without asking for anything more, Père Gaucher went back to his alembics as light as a swallow.

And in fact, from that time on, every evening after complines the officiating priest never failed to say—

"Let us pray for poor Père Gaucher, who is sacrificing his soul to the interest of the community. *Oremus Domine.*"*

And while over all these white cowls, prostrate in the shadow of the nave, the orison ran quivering like a little breeze over snow, at the other end of the convent, behind the flaming windows of the distillery Père Gaucher could be heard, singing at the top of his voice—

*O Lord, Let us pray.

Dans Paris il y a un Père Blanc,
Patatin, patatan, taraban, tarabin;
Dans Paris il y a un Père Blanc,
Qui fait danser des moinettes,
Trin, trin, trin, dans un jardin;
Qui fait danser, etc.*

Here the good curé stopped, filled with terror.
"Good heavens! Suppose my parishioners should hear me!"

Y

ERNEST DOWSON

(ENGLISH, 1867–1900)

Absinthia Taetra †

Green changed to white, emerald to an opal: nothing was changed.

The man let the water trickle gently into his glass, and as the green clouded, a mist fell away from his mind.

Then he drank opaline.

Memories and terrors beset him. The past tore after him like a panther and through the blackness of the present he saw the luminous tiger eyes of the things to be.

But he drank opaline.

And that obscure night of the soul, and the valley of humiliation, through which he stumbled were forgotten. He saw blue vistas of undiscovered countries, high prospects and a quiet, caressing sea. The past shed its perfume over him, to-day held his

* In Paris there is a White Father
  And so on and so forth;
  In Paris there is a White Father
  Who makes the little monks dance,
  Tra, la la, in the garden;
  Etc. . . .
† Poisonous Absinthe.

hand as it were a little child, and to-morrow shone like a white star: nothing was changed.

He drank opaline.

The man had known the obscure night of the soul, and lay even now in the valley of humiliation; and the tiger menace of the things to be was red in the skies. But for a little while he had forgotten.

Green changed to white, emerald to an opal: nothing was changed.

CHAPTER FIVE

DRUNK OR SOBER

CHARLES BAUDELAIRE

(1821-1867)

Get Drunk

One must always be drunk. Everything is there: It is the essential issue. To avoid that horrible burden of Time grabbing your shoulders and crushing you to the earth, you must get drunk without restraint.

But on what? On wine, poetry or virtue, whatever you fancy. But get drunk.

If sometimes, on palace walks, in the green grass of a ditch or the gloomy solitude of your room, you wake, and find that drunkenness is wearing off, ask the wind, the wave, the star, the bird, the clock, everything that flees, that laments, that turns, that sings, that speaks, ask what time it is; and the wind, the wave, the star, the bird, and the clock will answer you: "It is time to get drunk." To escape being the martyred slaves of time, be continually drunk. On wine, poetry or virtue, whatever you fancy.

—*Translated by John Digby*

CHARLES COTTON

(ENGLISH, 1630-1687)

Paraphras'd from Anacreon

The Earth with swallowing drunken showers
   Reels a perpetual round,
And with their Healths the Trees and Flowers
   Again drink up the Ground.

The Sea, of Liquor spuing full,
  The ambient Air doth sup,
And thirsty *Phoebus** at a pull
  Quaffs off the Ocean's cup.

When stagg'ring to a resting place,
  His bus'ness being done,
The Moon, with her pale platter face,
  Comes and drinks up the Sun.

Since Elements and Planets then
  Drink an eternal round,
'Tis much more proper sure for men
  Have better Liquor found.

Why may not I then, tell me pray,
Drink and be drunk as well as they?

PIERRE-JEAN de BÉRANGER

(FRENCH, 1780–1857)

The General Drinking Bout

LA GRANDE ORGIE.

A charm for every mother's son
Hath wine—then let it freely run,
      By the tun!
In floods of wine be Paris sunk;
We'll have your cross and crabbed old hunk
      Drunk!

---

* Apollo, the sun god.

Nay, there must be some pause
In the clutch of the laws;
Joyous Frenchmen, let's empty our cellars renowned!
Prosy censors in vain
From all wine may abstain;
They shall snuff but the fumes, and their brains shall spin round.

A charm for every mother's son
Hath wine—then let it freely run,
By the tun!
In floods of wine be Paris sunk;
We'll have your cross and crabbed old hunk
Drunk!

Authors grave, spouters chill,
Preachers harping on ill,
Ye whose hearers to slumber their senses resign,
Men of pamphlets, and men
Who on verse try the pen,
Come, exchange me your ink-horns for goblets of wine!
A charm for every mother's son
Hath wine—then let it freely run,
By the tun!
In floods of wine be Paris sunk;
We'll have your cross and crabbed old hunk
Drunk!

Mars, in taking his flight
From the din of the fight,
In our high-flavored wines would his thunderbolts steep:
From our arsenals roll.
O ye keepers, the whole
Of the barrels—for us—in which powder you keep.
A charm for every mother's son
Hath wine—then let it freely run,
By the tun!
In floods of wine be Paris sunk;

We'll have your cross and crabbed old hunk
Drunk!

We who hover where'er
Pretty rosebuds appear.
We'll muddle the doves that wing Venus's flight:
Birds to Cypris so dear,
Come, our noise never fear,
But, to drain the last drops on our glasses alight!
A charm for every mother's son
Hath wine—then let it freely run,
By the tun!
In floods of wine be Paris sunk;
We'll have your cross and crabbed old hunk
Drunk!

Gold is ten times too heavy;
A rollicking bevy
Of topers, who lush for their mistresses' sake,
On the whole will aver
That this glass they prefer
To the metal that fools into diadems make.
A charm for every mother's son
Hath wine—then let it freely run,
By the tun!
In floods of wine be Paris sunk;
We'll have your cross and crabbed old hunk
Drunk!

Charming children of mothers
Whose common sense smothers
The nonsense of sentiments grand as you please,
Sons of ours, brisk and plump,
Into being shall jump,
'Mid the wine-pots, their faces besmeared with the lees.
A charm for every mother's son
Hath wine—then let it freely run,
By the tun!

In floods of wine be Paris sunk;
We'll have your cross and crabbed old hunk
                    Drunk!

Then to honors a truce;
Let them cease to seduce:
We're at length truly happy—what, ho, for our signs!
    Jolly dogs, every king
    To his bottle shall cling,
And the bay-tree shall serve but to prop up our vines.
    A charm for every mother's son
    Hath wine—then let it freely run,
                    By the tun!
    In floods of wine be Paris sunk;
    We'll have your cross and crabbed old hunk
                    Drunk!

Reason, Reason, good bye!
On this spot where we lie,
Bowing down before Bacchus whose praise is our theme,
    Clad in purple or tatters,
    All friends for such matters,
We'll drop off to sleep, and of vintages dream!
    A charm for every mother's son
    Hath wine—then let it freely run,
                    By the tun!
    In floods of wine be Paris sunk;
    We'll have your cross and crabbed old hunk
                    Drunk!

♟

SEBASTIAN BRANT

(G E R M A N , 1 4 5 7 ? - 1 5 2 1)

From *The Ship of Fools*

OF GLUTTONY AND FEASTING

He shoes a fool in every wise
Who day and night forever hies
From feast to feast to fill his paunch
And make his figure round and staunch,
As though his mission he were filling
By drinking too much wine and swilling
And bringing hoar-frost o'er the grape.
Into the fool's ship toss the ape,
He kills all reason, is not sage,
And will regret it in old age.
His head and hands will ever shake,
his life a speedy end may take,
For wine's a very harmful thing,
A man shows no sound reasoning
Who only drinks for sordid ends,
A drunken man neglects his friends
And knows no prudent moderation,
And drinking leads to fornication;
It oft induces grave offense,
A wise man drinks with common sense.
For wine old Noah cared no whit,*

---

*The biblical *exempla* that follow were common in medieval preaching against drunkenness: Gen. 9:20; Gen. 19:33; Matt. 14; Exod. 32; Isa. 5:11; I Kings 20:16; Luke 16:19; Prov. 23:34; Prov. 23:31-2. Like his contemporaries Brant also draws on moral examples from ancient history and philosophy: King Cyrus from Herodotus, Alexander's murder of Cleitus from Plutarch, and the wisdom of Seneca. Chaucer uses many of the same references in "The Pardoner's Tale."

Although he found and planted it;
By wine Loth twice to sin was led,
Through wine the Baptist lost his head,
Through wine a wise man comes to prate
And set a fool's cap on his pate;
When Israelites were drunk with wine
And glutted full like silly swine,
They gamboled then in highest glee
And had to dance in revelry.
To Aaron's sons did God decree
That abstinent and chaste they be
And that to wine they should not turn,
But this decree the priests would spurn.
King Holofernes too when drunk,
He had his head cut off his trunk;
To feasts Tomyris had recourse
When old King Cyrus she would force;
Wine caused the fall of Ben-hadad,
Deprived was he of all he had;
When Alexander played the sot
His honor, virtue he forgot
And practiced deeds in drunkenness
That presently brought sad distress.
The rich man reveled once so well
That on the morn he ate in hell.
Man would not be a slave, in fine,
If he disowned the demon wine:
Are wine and sumptuous food your itch?
You'll not be happy, nor get rich.
Woe's him and woe's his father too,
He'll have misfortunes not a few
Who always gorges like a beast
Proposing toasts at every feast,
And would with others glasses clink;
The man whose joy is endless drink
Is like a man who falls asleep
Defenseless on the ocean deep;
Thus they who drink and e'er are gay,
Carousing, toping night and day:

If he's their friend, the generous host
Brings veal galore, a cow almost,
And gives them almonds, figs, and rice,
The bill, alas, is writ on ice.
Some men would be intelligent
From wine if wisdom e'er it lent,
Who cool their throats with rich libation.
Friend drinks to friend without cessation:
"I drink to you." "Here's happy days!"
"This cup for you." "This yours!" he says;
"I'll toast you till we both are filled!"
Thus speak the men of folly's guild.
Upset the glass, the drinker too,
A rope around his neck would do
Him better far than wild carousing
And naught but foolishness arousing,
That ancient Seneca did flay
In books that still are read today,
Which say one pays a drunken man
More heed than many a sober man,
And how an honor high 'tis rated
By wine to be intoxicated;
I censure those who tipple beer,
A keg of it per man, I hear,
Becoming so inebriate
That with them one could ope a gate.
A fool shows no consideration,
A wise man drinks with moderation,
Feels better, illness too defies,
Than one imbibing bucketwise.
The wine, 'tis true, our thirst will slake
But later stabs one like a snake,
And poison through the veins will pour,
As Basiliscus found of yore.

—*Translated by Edwin H. Zeydel*

♟

KENKŌ YOSHIDA

(J A P A N E S E ,   1 2 8 3 – 1 3 5 0)

From *The Tsurezuregusa,* 175

There are many things in the world I cannot understand. I cannot imagine why people find it so enjoyable to press liquor on you the first thing, on every occasion, and force you to drink it. The drinker's face grimaces as if with unbearable distress, and he looks for a chance to get rid of the drink and escape unobserved, only to be stopped and senselessly forced to drink more. As a result, even dignified men suddenly turn into lunatics and behave idiotically, and men in the prime of health act like patients afflicted with grave illnesses and collapse unconscious before one's eyes. What a scandalous way to spend a day of celebration! The victim's head aches even the following day, and he lies abed, groaning, unable to eat, unable to recall what happened the day before, as if everything had taken place in a previous incarnation. He neglects important duties, both public and private, and the result is disaster. It is cruel and a breach of courtesy to oblige a man to undergo such experiences. Moreover, will not the man who has been put through this ordeal feel bitter and resentful towards his tormentors? If it were reported that such a custom, unknown among ourselves, existed in some foreign country, we should certainly find it peculiar and even incredible.

I find this practice distressing to observe even in strangers. A man whose thoughtful manner had seemed attractive laughs and shouts uncontrollably; he chatters interminably, his court cap askew, the cords of his cloak undone, the skirts of his kimono rolled up to his shins, presenting so disreputable a picture that he is unrecognizable as his usual self. A woman will brush the hair away from her forehead and brazenly lift up her face with a roar of laughter. She clings to a man's hand as he holds a saké cup, and if badly bred she will push appetizers into the mouth of her companion, or her own, a disgraceful sight. Some men shout at the top of their lungs, singing and dancing, each to his own tune. Sometimes an old priest, invited at the behest of a distinguished guest, strips to the waist, revealing grimy, sallow skin, and twists his body in a manner so revolting that even those watching with amusement are nauseated. Some drone on about their achievements, boring their listeners; others weep drunkenly. People of the lower classes swear at one another and quarrel in a shocking and frightening manner; after various shameful and wretched antics they end up by grabbing things they have been refused, or falling from the verandah (or from a horse or a carriage) and injuring themselves. Or, if they are not sufficiently important to ride, they stagger along the main thoroughfares and perform various unmentionable acts before earthen walls or at people's gates. It is most upsetting to see an old priest in his shawl leaning on the shoulder of a boy and staggering along, mumbling something incomprehensible.

If such behavior were of benefit either in this world or the next, there might be some excuse. It is, however, the source of numerous calamities in this world, destroying fortunes and inviting sickness. They call liquor the chief of all medicines, but it is, in fact, the origin of all sicknesses. Liquor makes you forget your unhappiness, we are told, but when a man is drunk he may remember even his past griefs and weep over them. As for the future life, liquor deprives a man of his wisdom and consumes his good actions like fire; he therefore increases the burden of sin, violates many commandments and, in the end, drops into hell. Buddha taught that a man who takes liquor and forces another to drink will be reborn five hundred times without hands.

Though liquor is as loathsome as I have described it, there naturally are some occasions when it is hard to dispense with. On a moonlit night, a morning after a snowfall, or under the cherry blossoms, it adds to our pleasure if, while chatting at our ease, we bring forth the wine cups. Liquor is cheering on days when we are bored, or when a friend pays an unexpected visit. It is exceedingly agreeable too when you are offered cakes and wine most elegantly from behind a screen of state by a person of quality you

do not know especially well. In winter it is delightful to sit opposite an intimate friend in a small room, toasting something to eat
over the fire, and to drink deeply together. It is pleasant also when
stopping briefly on a journey, or picnicking in the countryside, to
sit drinking on the grass, saying all the while, "I wish we had
something to eat with this saké." It is amusing when a man who
hates liquor has been made to drink a little. How pleasing it is,
again, when some distinguished man deigns to say, "Have another. Your cup looks a little empty." I am happy when some man
I have wanted to make my friend is fond of liquor, and we are
soon on intimate terms.

Despite all I have said, a drinker is amusing, and his offense is
pardonable. It happens sometimes that a guest who has slept late
in the morning is awakened by his host flinging open the sliding
doors. The startled guest, his face still dazed by sleep, pokes out
his head with its thin topknot and, not stopping to put on his
clothes, carries them off in his arms, trailing some behind as he
flees. It is an amusing and appropriate finale to the drinking party
to catch a glimpse of the skinny, hairy shanks he reveals from behind as he lifts his skirts in flight.

—*Translated by Donald Keene*

♟

HOWARD MOSS

(AMERICAN, 1922–1987)

The Writer at the End of the Bar

Dated at forty-five? Not yet,
You sly, old Phoenix in leather boots.
Infancy, with its double takes,
Is your seismograph, and its earthquakes.

Tell me, what are you writing now—
My Damaged Nerves: The Great Shakes?

Dumbshows count, not words. You are
Silent, maimed. Slumped at a bar,
What's left to learn in its mirrored length
That isn't already mirrored at length
In your life? Is it true you are
Writing *Oh, God, Give Me Strength*—

The Weak Shall Inherit Each Other?
They'd better; they don't inherit the earth.
Sometimes they don't—or can't or won't—
Even inherit each other. So
Why revise *The Prodigal Sons:*
No Deposits and No Returns?

Get up, get out. Night's dark and cheap.
Hell is a place that's never filled.
Maybe that's where you'll sleep tonight,
Writing in dreams, as the sleeve unravels
Your latest non-best-seller,
Down in the Dumps and Other Travels.

SAMUEL BUTLER

(ENGLISH, 1612–1680)

From *Characters*

A SOT

Has found out a Way to renew, not only his Youth, but his Childhood, by being stewed, like old *Æson*, in Liquor;* much better
than the *Virtuoso's* Way of making old Dogs young again: for he

* Jason's father rejuvenated by Medea's injection of herbal juice.

is a Child again at second hand, never the worse for the Wearing, but as purely fresh, simple, and weak, as he was at first. He has stupify'd his Senses by living in a moist Climate according to the Poet—*B[oe]otum in crasso jurares aëre natum.** He measures his Time by Glasses of Wine, as the Ancients did by Water-Glasses; and as *Hermes Trismegistus*† is said to have kept the first Accompt of Hours by the pissing of a Beast dedicated to *Serapis*,‡ he revives that Custom in his own Practice, and observes it punctually in passing his Time. He is like a Statue placed in a moist Air; all the Lineaments of Humanity are mouldered away, and there is nothing left of him but a rude Lump of the Shape of a Man, and no one part entire. He has drowned himself in a But of Wine, as the Duke of *Clarence*§ was served by his Brother. He has washed down his Soul and pist it out; and lives now only by the Spirit of Wine or Brandy, or by an Extract drawn off his Stomach. He has swallowed his Humanity, and drunk himself into a Beast, as if he had pledged *Madam Circe*,‖ and done her Right. He is drowned in a Glass like a Fly, beyond the Cure of Crums of Bread, or the Sun Beams. He is like a Spring-Tide; when he is drunk to his high-Water-Mark he swells and looks big, runs against the Stream, and overflows every Thing that stands in his Way; but when the Drink within him is at an Ebb, he shrinks within his Banks, and falls so low and shallow, that Cattle may pass, over him. He governs all his Actions by the Drink within him, as a *Quaker* does by the Light within him; has a different Humour for every Nick his Drink rises to, like the Degrees of the Weatherglass, and proceeds from Ribaldry and Bawdery to Politics, Religion, and Quarreling, until it is at the Top, and then it is the Dog-Days with him; from whence he falls down again, until his Liquor is at the Bottom, and then he lyes quiet, and is frozen up.

*You would swear he was born in the heavy Beotian air.
†Name applied to the authors of diverse cryptic writings of a philosophical and scientific nature.
‡Baboon.
§*Richard III*, I, iv.
‖*Odyssey* X.

Y

JOHN DIGBY

(1 9 3 8 –)

Incident at the Gaumont

"You ain't wearing that silly hat again are you?" questioned Flossie who appeared behind Gertie as she was steadying a bright colored hat on her head in front of a mirror.

"Of course, wide brims suit my full face," answered Gertie.

"Talk about a blood pudding with its top ain't the word."

"Thank you for being so kind and frank," Gertie said with a tone of irony in her voice.

"Well, I mean you don't want to make a right fool of yourself do you."

"I wear hats because they suit me and who knows I might meet Mr. Right."

"I hate to say it, but I'll bet you'll only meet Mr. Wrong in that hat," Flossie answered, straightening the seam of her nylon stocking.

"Whatever you say, I'm still going to wear it." Gertie was adamant.

"I can't see how you or me, come to that matter, will meet either Mr. Right or Mr. Wrong at the pictures at eight o'clock at night," offered Flossie.

The hat in question was a bright cherry-red object precariously balanced on Gertie's piled-up hair.

"There," she said, "it's fixed."

"I do hope the person behind you don't ask you to remove it," said Flossie. "It's so embarrassing."

"O do shut up about my hat, and put on a pair of flats as we have a bit of a walk before we reach the Gaumont."

The two girls were at last ready. They considered it prudent to leave a little earlier as the film showing that night had Robert Young as the star. They thought there might be a queue and it was wise to arrive a little before time in order to be first in the cinema so that they could pick their seats, not too near the front of the screen.

It was a long walk to the cinema and both girls were happy that they had chosen flats that night.

"We can take them off in the pictures," Flossie said.

"Take what off?" questioned Gertie. .

"Our shoes of course, we better not forget though."

"O yes, but our feet might swell," said Gertie.

"I'll take that chance."

They arrived at the cinema and found it deserted.

"You think we've come the wrong day?"

"More like the wrong time," answered Gertie. She peered through the glass doors in order to read the times of the program.

"It ain't eight o'clock, it's nine o'clock the film starts," Flossie said rather sadly.

"I do hate going in half way through the picture," said Gertie.

"So do I. And I want to go to the toilet," answered Flossie.

"Trust you. Why didn't you go before we left home?"

"Never felt like it," answered Flossie.

"Ask if you can use the one inside."

"I don't fancy that, and anyway the toilet's on the right inside the cinema and I will have to walk past the screen to get there."

"Well I'm not going home again. You shall have to go in the car park behind the bushes," said Gertie in a forbidding voice.

"Say if someone sees me?"

"Who's going to watch you? Mr. Right?" asked Gertie.

It was a long walk back home, and if they attempted it they well might miss the opening of the film. Flossie paused for a second and considered her plan of action. Finally she made up her mind.

"All right, I'll go. Will you watch for me?"

"Yes, yes," answered Gertie impatiently.

Both girls walked to the back of the cinema. Gertie stood with her back to the car park while Flossie wandered down into the darkness, hesitantly, endeavoring to find a good clump of bushes in which to hide herself. It was all over in a couple of seconds. Flossie came back running.

"Seen a ghost?" asked Gertie.

"Seen a ghost," repeated Flossie, "I've seen a dead man. There," she pointed at the back of the car park.

"Get away with you," Gertie said with a tone of disbelief in her voice.

"Come and see. Just you come and see," said Flossie.

The two girls walked back to where the dead man lay behind the bushes.

"I ain't never seen a dead man before," said Gertie as both girls peered down at the crumpled body lying in a grotesque position.

"It gives me the right willies," said Flossie, "and to think that he was there when I was. . . . What are we going to do?"

"Inform the police," Gertie said in no uncertain tone.

Both girls walked briskly around to the front of the cinema where they interrupted the attendant's conversation with the ticket-sale girl.

"We want to see the manager," Flossie said. Her voice was sharp.

"All right girls. I'll get him, just wait here." The attendant looked at the two girls with a puzzled look, but as Flossie sounded so urgent he never questioned their sincerity.

"We might get our names and pictures in the papers," Gertie said.

Flossie said nothing. The girls waited and the manager appeared. He was a young man with a pencil-slim moustache. His hair was heavily greased and combed back with a perfect parting. Gertie thought he looked more like a third-rate dance-hall instructor than a cinema manager.

"We've found a dead man in your car park," Gertie blurted breathlessly.

"A dead man. In the car park?" The manager repeated his words as if he were speaking to a couple of infants who were having difficulty in mastering their native tongue.

"Yes," said Flossie, "a dead man."

"Really dead," echoed Gertie.

The young man straightened himself to attention as if he were in the army and acting under orders, or some invisible strings pulled him suddenly together.

"You better show me," he announced. His voice grew serious and his face turned earnest.

"We'll point him out to you," Gertie said.

The manager walked smartly around the back of the cinema to the car park with the two girls trailing behind him.

"I hope this isn't some kind of joke," he said suddenly, without looking at the girls.

"There, down there," said Flossie, pointing to the dark at the end of the car park. "Just near the tall bushes there, before you reach the broken fence. That wire fence on the left."

He walked down into the darkness. He was gone less than a couple of minutes. He was visibly shaken when he returned.

"You two young ladies better come back with me to the office," he said.

"Well. Were we right?" asked Gertie. The manager never answered her question.

"You two girls sit here while I make a phone call." He closed the door behind him and left the girls alone in his small office.

"I bet we'll get our names in the papers, might even get our photos taken," Gertie said, straightening her hat.

"Fancy finding a dead man," Flossie said, muttering to herself.

"Doesn't explain what we were doing at the back of the car park though."

"O dear, I never thought of that. It could be a bit embarrassing," replied Flossie.

"If the worst comes to the worst we shall have to swallow our pride."

"I guess so," said Flossie, almost giving way to a slight blush.

It seemed an interminable period before the door opened and in walked two police constables with a plainclothes detective. The cinema manager walked in, or rather looked in, sheepishly, and then left in a hurry. The two policemen stood at either side of the door as if they were preventing any possible escape. The detective, who looked more like a school master than anything else, walked to the middle of the room and stood in front of the girls. He looked at them and remained silent, as if searching for words. The two girls appeared excited; each one had her own involved story to tell. Both were ready to welcome questions—all questions that the detective might care to ask. They had both forgotten the reason why Flossie had gone to the back of the car park.

"Which one found the man?" the detective asked, smiling.

"Will we get our names in the papers?" Gertie asked.

"Names. Newspapers?" The question threw the detective off balance momentarily.

"He's dead ain't he?" Flossie asked with a tinge of pity in her voice.

"Dead? Yes, he's dead all right," offered the detective.

"Ain't we good enough for the papers," replied Gertie.

The detective ignored Gertie's question and a wide grin appeared across his face. He looked at the two girls with a mixture of contempt and pity.

"He's dead all right girls. Dead drunk with half a bottle of whiskey inside him and the other half inside his coat pocket."

The two policemen laughed, and even the detective, despite his efforts to remain serious, started to laugh also.

Outside the cinema the girls appeared rather disappointed about their adventure that failed to materialize.

"You and your dead bodies," Gertie said.

"You're just as bad. 'Will we get our names and pictures in the papers,'" Flossie replied.

It was too late to see the film now. They had missed Robert Young. They started to walk home. A light rain was beginning to fall. They walked in silence. Finally Gertie spoke.

"To think that we missed Robert Young because of a drunk!"

Even if they were a little crestfallen the two girls couldn't help laughing.

☙

LOUIS MACNEICE

(1 9 0 7 – 1 9 6 3)

The Drunkard

His last train home is Purgatory in reverse,
A spiral back into time and down towards Hell
Clutching a quizzical strap where wraiths of faces

Contract, expand, revolve, impinge; disperse
On a sickly wind which drives all wraiths pell-mell
Through tunnels to their appointed, separate places.

And he is separate too, who had but now ascended
Into the panarchy of created things
Wearing his halo cocked, full of good will
That need not be implemented; time stood still
As the false coin rang and the four walls had wings
And instantly the Natural Man was mended.

Instantly and it would be permanently
God was uttered in words and gulped in gin,
The barmaid was a Madonna, the adoration
Of the coalman's breath was myrrh, the world was We
And pissing under the stars an act of creation
While the low hills lay purring round the inn.

Such was the absolute moment, to be displaced
By moments; the clock takes over—time to descend
Where Time will brief us, briefed himself to oppress
The man who looks and finds Man human and not his friend
And whose tongue feels around and around but cannot taste
That hour-gone sacrament of drunkenness.

ERNEST HEMINGWAY

(AMERICAN , 1 8 9 9 – 1 9 6 1)

A Clean, Well-Lighted Place

It was late and every one had left the café except an old man who
sat in the shadow the leaves of the tree made against the electric
light. In the day time the street was dusty, but at night the dew

settled the dust and the old man liked to sit late because he was deaf and now at night it was quiet and he felt the difference. The two waiters inside the café knew that the old man was a little drunk, and while he was a good client they knew that if he became too drunk he would leave without paying, so they kept watch on him.

"Last week he tried to commit suicide," one waiter said.

"Why?"

"He was in despair."

"What about?"

"Nothing."

"How do you know it was nothing?"

"He has plenty of money."

They sat together at a table that was close against the wall near the door of the café and looked at the terrace where the tables were all empty except where the old man sat in the shadow of the leaves of the tree that moved slightly in the wind. A girl and a soldier went by in the street. The street light shone on the brass number on his collar. The girl wore no head covering and hurried beside him.

"The guard will pick him up," one waiter said.

"What does it matter if he gets what he's after?"

"He had better get off the street now. The guard will get him. They went by five minutes ago."

The old man sitting in the shadow rapped on his saucer with his glass. The younger waiter went over to him.

"What do you want?"

The old man looked at him. "Another brandy," he said.

"You'll be drunk," the waiter said. The old man looked at him. The waiter went away.

"He'll stay all night," he said to his colleague. "I'm sleepy now. I never get into bed before three o'clock. He should have killed himself last week."

The waiter took the brandy bottle and another saucer from the counter inside the café and marched out to the old man's table. He put down the saucer and poured the glass full of brandy.

"You should have killed yourself last week," he said to the deaf man. The old man motioned with his finger. "A little more," he said. The waiter poured on into the glass so that the brandy slopped over and ran down the stem into the top saucer of the pile. "Thank you," the old man said. The waiter took the bottle back inside the café. He sat down at the table with his colleague again.

"He's drunk now," he said.

"He's drunk every night."

"What did he want to kill himself for?"

"How should I know."

"How did he do it?"

"He hung himself with a rope."

"Who cut him down?"

"His niece."

"Why did they do it?"

"Fear for his soul."

"How much money has he got?"

"He's got plenty."

"He must be eighty years old."

"Anyway I should say he was eighty."

"I wish he would go home. I never get to bed before three o'clock. What kind of hour is that to go to bed?"

"He stays up because he likes it."

"He's lonely. I'm not lonely. I have a wife waiting in bed for me."

"He had a wife once too."

"A wife would be no good to him now."

"You can't tell. He might be better with a wife."

"His niece looks after him."

"I know. You said she cut him down."

"I wouldn't want to be that old. An old man is a nasty thing."

"Not always. This old man is clean. He drinks without spilling. Even now, drunk. Look at him."

"I don't want to look at him. I wish he would go home. He has no regard for those who must work."

The old man looked from his glass across the square, then over at the waiters.

"Another brandy," he said, pointing to his glass. The waiter who was in a hurry came over.

"Finished," he said, speaking with that omission of syntax stupid people employ when talking to drunken people or foreigners. "No more tonight. Close now."

"Another," said the old man.

"No. Finished." The waiter wiped the edge of the table with a towel and shook his head.

The old man stood up, slowly counted the saucers, took a leather coin purse from his pocket and paid for the drinks, leaving half a peseta tip.

The waiter watched him go down the street, a very old man walking unsteadily but with dignity.

"Why didn't you let him stay and drink?" the unhurried waiter asked. They were putting up the shutters. "It is not half-past two."

"I want to go home to bed."

"What is an hour?"

"More to me than to him."

"An hour is the same."

"You talk like an old man yourself. He can buy a bottle and drink at home."

"It's not the same."

"No, it is not," agreed the waiter with a wife. He did not wish to be unjust. He was only in a hurry.

"And you? You have no fear of going home before your usual hour?"

"Are you trying to insult me?"

"No, hombre, only to make a joke."

"No," the waiter who was in a hurry said, rising from pulling down the metal shutters. "I have confidence. I am all confidence."

"You have youth, confidence, and a job," the older waiter said. "You have everything."

"And what do you lack?"

"Everything but work."

"You have everything I have."

"No. I have never had confidence and I am not young."

"Come on. Stop talking nonsense and lock up."

"I am of those who like to stay late at the café," the older waiter said. "With all those who do not want to go to bed. With all those who need a light for the night."

"I want to go home and into bed."

"We are of two different kinds," the older waiter said. He was now dressed to go home. "It is not only a question of youth and confidence although those things are very beautiful. Each night I am reluctant to close up because there may be some one who needs the café."

"Hombre, there are bodegas* open all night long."

"You do not understand. This is a clean and pleasant café. It is well lighted. The light is very good and also, now, there are shadows of the leaves."

"Good night," said the younger waiter.

"Good night," the other said. Turning off the electric light he continued the conversation with himself. It is the light of course but it is necessary that the place be clean and pleasant. You do not want music. Certainly you do not want music. Nor can you stand before a bar with dignity although that is all that is provided for these hours. What did he fear? It was not fear or dread. It was a nothing that he knew too well. It was all a nothing and a man was nothing too. It was only that and light was all it needed and a certain cleanness and order. Some lived in it and never felt it but he knew it all was nada y pues nada y nada y pues nada.† Our nada who art in nada, nada be thy name thy kingdom nada thy will be nada in nada as it is in nada. Give us this nada our daily nada and nada us our nada as we nada our nadas and nada us not into nada but deliver us from nada; pues nada. Hail nothing full of nothing, nothing is with thee. He smiled and stood before a bar with a shining steam pressure coffee machine.

"What's yours?" asked the barman.

"Nada."

"Otro loco más,"‡ said the barman and turned away.

"A little cup," said the waiter.

The barman poured it for him.

"The light is very bright and pleasant but the bar is unpolished," the waiter said.

The barman looked at him but did not answer. It was too late at night for conversation.

"You want another copita?" the barman asked.

"No, thank you," said the waiter and went out. He disliked bars

* Groceries.
† An idiomatic accumulation of nothing.
‡ The other is crazier.

and bodegas. A clean, well-lighted café was a very different thing. Now, without thinking further, he would go home to his room. He would lie in the bed and finally, with daylight, he would go to sleep. After all, he said to himself, it is probably only insomnia. Many must have it.

Y

RICHMOND LATTIMORE

(AMERICAN, 1906–1984)

Drunken Old Solipsists in a Bar

In their own cool gray alcoholic world
sealed from the sun at any time of day,
you find this circle of old heads. A glass
that fills and drains and fills again with gold
sits before each, to tranquilize the spirit
and burn slow fires in the stupescent brain.

The gray bar is the inside of a brain
selfgrown and shuttered on the outward world,
where, clammed in its own fumy smell, the spirit
rapt in the microcosm of its day
spells out the cloistered hours and guzzles gold
from the inverted barrel of the glass;

as if, gray in a cave or belled in glass,
this bar, the unique inside of the brain,
trimmed with brown wood and bottles painted gold,
subsisted as the all and only world,
with no outside, no windows, and no day,
its walls, all body grown upon the spirit;

and there, reliving glory in strong spirits
as in the flattery of a magic glass,
they know themselves as on a younger day
with April lyrics singing in the brain,
the past recaptured to a rainbow world
drenched in fond sunshine and philosopher's gold.

The catfoot barkeep doles his bottled gold,
and television, ectoplasmic spirit,
gray-glimmering ghost of the external world,
gibbers and mimes behind its convex glass,
with incantations feeds and stills the brain
on distillations filmed from the live day.

Then doors open, and into dazzling day
of summer afternoon with all its gold,
old knees, jerked by the strings of a half-blind brain,
float them toward home like disembodied spirits,
with faces stuffed and set, and eyes like glass,
and one still grin to give to all the world.

All through the drunken day this larval spirit
builds his bright palaces of gold and glass
in the domed brain which forms his private world.

CARL VAN VECHTEN

(AMERICAN , 1 8 8 0 – 1 9 6 4)

From *Parties*, Chapter 6

In Rilda's bedroom, David lay nude on his belly in a splash of
sunlight which mottled his body until it resembled a painting by
Monet. Idly kicking his heels in the air, he played with a flexible

brass fish while Rilda in a lace dressing-gown sipped her coffee and smoked a cigarette. A copy of *The New York Times* was spread out, unread, at her feet. Outside there was a prodigious honking of motor-horns and the hammering of a riveter as it flattened the headless ends of bolts in a steel construction across the way. Through the window men, poised perilously at a high altitude on the cross-beams, might be observed catching these bolts as they were tossed molten from the furnace. David glanced upwards occasionally to watch these operations.

Of course, I haven't lost everything, he said after a long silence, but it will make quite a dent in our income. I wouldn't have lost anything if my broker hadn't asked my advice. I was a fool to give it to him.

Rilda regarded her husband with an expression denoting complete infatuation.

I wonder why I adore you so, David: you're really such a bastard.

I suspect it's because I'm so swell, David replied complacently.

You *are* swell, David, when you are sober.

Snap out of it Rilda . . . David was scowling . . . You know damned well if you didn't love me when I was drunk, you wouldn't have much chance to love me at all. I love *you* when you are drunk, Rilda, and I think I'm a little jealous of that German.

What about Noma Ridge? Do I have to put up with that?

Don't be ridiculous, Rilda. Noma belongs to the town, to the country, in fact. She is really "America's sweetheart."

And Rosalie . . .

I can't think how you do it!

Yes, Rosalie. You certainly cannot imagine it gives me pleasure to dine with that bitch.

After all, you're never asked.

Whither thou goest . . .

Yes, we have to do that, don't we? David demanded, wriggling his great toes and wrinkling his nose. He was staring very hard at the painting of a farmhouse in the snow by Charles Burchfield,* as if he were seeing it for the first time . . . Aside from our booz-

---

*Midwestern American watercolorist and scene painter (1893–1967).

ing, it's the worst thing about us, our damned faithfulness to each other.

Rilda laughed, a little bitterly. Our damned faithfulness, as you call it, our "clean" fidelity, doesn't get us very far, she replied. We follow each other around in circles, loving and hating and wounding. We're both so sadistic. It's really too bad one of us isn't a masochist.

I *am* a masochist, David boasted. I love to have you hurt me.

You are certainly a liar, Rilda retorted. After a pause, she added, David, do you realize that this is the first time I've seen you alone in months?

David wriggled his great toes and watched the workmen through the window.

We are really too shy to be natural when we are alone together, he responded. We become self-conscious and talk the way they do in books—I mean in good books, of course! Running his fingers through his black curls, he inquired casually: Rilda, what *do* you see in that Siegfried person? I suspect it's his name, he added.

What do *you* see in Rosalie? Why do you spend all your time in Harlem with dope-addicts and bootleggers? I'll quit if you quit.

I don't want to quit, David replied grimly. Tired of the fish, he cast it aside.

Darling, I don't mind anything so long as I am with you, Rilda assured him, but when you float off on one of those long, vague, dangerous drunks when you don't know where you are or what you are or with whom you are, I worry.

You call up Hamish, David frowned.

I call up Hamish, she admitted.

And Donald . . .

And Donald.

You go to Rosalie's for dinner, where you are not invited, and the Gräfin's for cocktails and you behave outrageously with a young blond with the incredible name of Siegfried. *I do not like it.*

No more do I, you silly.

Rilda, what do you think of the Gräfin?

I think she is a swell person. I can't help liking the Gräfin. She is so simple and so direct and it's so wonderful of her to know what she really wants.

And to get it, too, David reflected aloud. I wonder if she is amusing us or we are amusing her.

Both, of course.

There was silence for a moment. Presently David began tentatively, and he was speaking very seriously, Rilda, do you know what I think?

What? Rising, she crossed the room and, seating herself on a cushion by his side, began to stroke his back.

He went on talking: I think it is time we separated.

Separated! . . . Slipping her hand through his hair, she gave his black curls a sudden tug . . . Why, David, I couldn't live without you!

Vos beaux yeux vont pleurer.* That's just the trouble. You know that song of Jimmie Durante's: I go my way and you go my way.

Parties! Parties! If I didn't go your way I'd never see you. Why won't you stay home occasionally?

I am home, but it's the same thing. We're shy and self-conscious, and faithful. It's this damned faithfulness that's the trouble, Rilda. We've got to get over this damned faithfulness. It's killing us. It's tearing us apart.

This "clean" fidelity. Parties! Parties!

Rilda, David announced with determination. I'm going somewhere. I think it will be London, he added, I'm so sick of seeing Englishmen.

It's no time of year to go to London. It's so foggy and cold and you're always undressing. You'll catch pneumonia and pleurisy . . .

And psittacosis? Not with the amount of Scotch I'll inhale.

You don't actually intend to go, David?

He sat up to face her, doubling his knees under him.

I *do* actually mean it, Rilda. I want to get away. Nobody we know does anything but drink in this crazy town. I'm bored. If you go with me there'll be the same strain, the same pull between us . . .

Rilda looked very cold and hard . . . I suppose you want to take Roy Fern, she said . . . She was playing with a nail-polisher.

---

* Your beautiful eyes are tearful.

David laughed at this. That funny little rotter! he replied. No, my dearest dear, I want to go alone. I want to get away . . .

from it all. She concluded his sentence for him. When did this plan occur to you?

Weeks ago.

At a party, I suppose.

Well, it might have been. Or in a speakeasy. Must have been, in fact. I suppose you will say I was drunk. I was. I do all my important thinking when I'm drunk because then my thinking expresses my feeling. I wouldn't dare make a decision like this when I'm sober: I wouldn't actually know whether I felt that way or not.

What exactly is it that you want to get away from? she asked him.

From you, my dear, or from *us,* from what it is that makes us hate and love and drink, from this intensity of "clean" fidelity, as you call it. I want to be actually unfaithful to you in feeling and imagination as well as physically, so that I can return to you free. Now I am your slave. I never make a move or commit an action without considering whether it will annoy you or not. I swear that the strongest sensation I experience when I look at another woman is to wonder what effect it will have on you. That's why I get drunk so often. That's why you get drunk so often. We get drunk to forget we belong to each other and when we are drunk we remember harder than ever. We waltz around and around like Japanese mice. Are you going to follow me to Rosalie's again tonight?

Good God, David, you can't go to that beastly woman's house again!

I promised I would.

Well, this time I won't follow you!

Brava! But it won't matter, Rilda, because you'll be there in my mind, or somewhere else with Siegfried. Don't you see what I mean, Rilda, my love? Don't you see what I mean? . . . Raising his hands to her shoulders he buried his face in the soft lace that covered her breast.

Of course, I see, David, and that is why I can't let you go. Suppose you enjoy your freedom, your mental infidelity and all the rest of it? Then you wouldn't come back at all. I couldn't bear that, David. What would I do without you?

Go on being outrageously hard and cynical with blond, mythological Germans, I suppose . . . David removed his head from Rilda's breast.

She was thoughtful.

I wouldn't do anything like that, she replied after a pause for reflection. Do you know, I don't think I'd see anybody at all.

In that case I'm sure I'd better go away.

David, if you go away I'll follow you . . .

Please, Rilda . . . He lay back on the blue carpet.

We might get a new set, was her next suggestion.

David's laughter momentarily drowned out the din of the riveting and the taxi horns.

Rilda, my pretty, there's no such thing as a set any more and you know it. Everybody goes everywhere. So that wouldn't help. . . . I can't stand the racket any longer. I mean that in every sense . . . He waved his hand in the direction of the street noises . . . I'm going to embark on something large and important like the Bremen or the Majestic.

You sweet swine, you know how I hate the sea at this time of year.

Rilda . . . David was still very solemn . . . you are not going with me. I've got to go on my own for awhile, till I work this thing out.

Scrutinizing his face intently, she demanded, Do you really mean that, David?

Yes, he replied, but some inner necessity impelled him to go on. Do you remember, dearest Rilda, the night we spent in Granada, the flowing water, the nightingales, the green of the trees, and the blind musicians playing the music of Manuel de Falla* in the garden?

Of course, David, and the blind gipsy who plucked his guitar strings and wailed of love in the street in front of the hotel, and the dog that barked all night, and the tinkling bells on the sturdy burros, and the flamencas with the red carnations stuck straight upright in their hair. I don't think I've ever been happier.

---

* Spanish classical composer (1876–1946).

I want to try to recapture all that, Rilda, all that we were to each other and the sort of things we did then. I want to get all that back, Rilda.

She stooped to kiss his eyes.

I wonder . . . she began.

What?

If you can get it back by leaving me.

Dear Rilda, I am firm. I shall go away alone.

Rising from the cushion, she recrossed the room to stand by the window. The sound of the riveting had become deafening and persistent.

You're sure it isn't Noma or Rosalie . . . or Roy? she inquired, not without malice.

Almost sure, Rilda . . . David was very gentle . . . As sure as I can be . . . I wonder I've got the nerve to leave you. If I stay here any longer I won't have. I won't have a bit of character left. I'll just be a drunken jelly-fish swimming around my old ideal of you and polluting it. Yet, if I leave you, I'll worry furiously about this Siegfried person. You really don't know how jealous I can be.

I know how jealous *I* can be, Rilda replied. You don't think I would dine uninvited with that bitch Rosalie Keith unless I was pretty damned jealous, do you?

David lay on his back and stretched his arms.

We're shattered, Rilda, he said. What we need is a drink. It's pretty near lunch time. We've had too much sober sleep. We're not used to it. The sun's too bright. That riveting is like life in the trenches. What we need is a coupla sidecars.

Want to go into the bar?

No, there's sure to be somebody there. I don't feel up to the others yet.

Tossing a square of crimson velvet across her husband's recumbent figure, Rilda rang the bell.

A Negro maid opened the door.

Have we any cointreau, Edith? Rilda inquired.

I think so. I'll just have a look, the girl replied.

If there is, make some sidecars. Otherwise use the gin with five fruits, or make something out of absinthe and corn-meal. Is there any one in the bar, Edith?

Yes, the maid answered, before she left the room, Mrs. Fly and Mr. Butcher.

We'll stay here, David reaffirmed, adding, as the telephone bell tinkled, I'm not at home to anybody till I've had at least one drink.

Rilda lifted the receiver.

Hello . . . Who? . . . Oh yes . . . I don't think so . . . No, I can't today . . . Well, last night was different. Last nights are always different . . . No, I'm not going to Rosalie's . . . If you like.

She replaced the receiver.

The bloody boche, I suppose, said David, kicking his heels in the air.

Well, I'm not seeing him.

I know all about *that,* David announced. You told him to call up later. That means that after cocktails, after a few whiskies and sodas, or a bottle or two of champagne, when you are a little tighter, you will see him.

And what about you? she demanded.

I've just been admitting it, he went on. We're swine, filthy swine, and we are Japanese mice, and we are polar bears walking from one end of our cage to the other, to and fro, to and fro, all day, all week, all month, for ever to eternity. We'll be drunk pretty soon and then I'll be off to Donald's to get drunker and you'll be off with Siegfried and get drunker and we'll go to a lot of cocktail parties and then we'll all turn up for dinner at Rosalie's where you are never invited. She won't want you, and I shall hate you, but Siegfried will want you. And we'll get drunker and drunker and drift about night clubs so drunk that we won't know where we are, and then we'll go to Harlem and stay up all night and go to bed late tomorrow morning and wake up and begin it all over again.

Parties, sighed Rilda. Parties!

Edith returned with the cocktail shaker and glasses on a tray.

There was plenty of cointreau, Mrs. Westlake, she announced.

So you made sidecars, David suggested.

So I made sidecars, Edith admitted and quietly left the room.

Filling two glasses, Rilda handed one to David. Draining it at one gulp, he passed it back to be refilled.

I am going away, Rilda, he said. Then he added, *Alone.*

⚱

JAMES WRIGHT

(AMERICAN , 1 9 2 7 – 1 9 8 0)

Two Hangovers

NUMBER ONE
I slouch in bed.
Beyond the streaked trees of my window,
All groves are bare.
Locusts and poplars change to unmarried women
Sorting slate from anthracite
Between railroad ties:
The yellow-bearded winter of the depression
Is still alive somewhere, an old man
Counting his collection of bottle caps
In a tarpaper shack under the cold trees
Of my grave.

I still feel half drunk,
And all those old women beyond my window
Are hunching toward the graveyard.

Drunk, mumbling Hungarian,
The sun staggers in,
And his big stupid face pitches
Into the stove.
For two hours I have been dreaming
Of green butterflies searching for diamonds
In coal seams;
And children chasing each other for a game
Through the hills of fresh graves.
But the sun has come home drunk from the sea,
And a sparrow outside
Sings of the Hanna Coal Co. and the dead moon.

The filaments of cold light bulbs tremble
In music like delicate birds.
Ah, turn it off.

NUMBER TWO: I TRY TO WAKEN AND GREET THE
WORLD ONCE AGAIN
In a pine tree,
A few yards away from my window sill,
A brilliant blue jay is springing up and down, up and down,
On a branch.
I laugh, as I see him abandon himself
To entire delight, for he knows as well as I do
That the branch will not break.

▼

WILLIAM SHAKESPEARE

(1 5 6 4 – 1 6 1 6)

From *Macbeth*, II, iii

MACDUFF. Was it so late, friend, ere you went to bed, that
you do lie so late?

PORTER. Faith, sir, we were carousing till the second cock;
and drink, sir, is a great provoker of three things.

MACDUFF. What three things does drink especially provoke?

PORTER. Marry, sir, nose-painting, sleep, and urine. Lechery,
sir, it provokes, and unprovokes; it provokes the desire, but it
takes away the performance; therefore, much drink may be said to
be an equivocator with lechery: it makes him, and it mars him; it
sets him on, and it takes him off; it persuades him, and disheartens
him; makes him stand to, and not stand to; in conclusion, equiv-
ocates him in a sleep, and, giving him the lie, leaves him.

MACDUFF. I believe drink gave thee the lie last night.

PORTER. That it did, sir, i' the very throat on me. But I requited him for his lie; and, I think, being too strong for him, though he took up my legs sometime, yet I made a shift to cast him.

Y

ALPHONSE DAUDET

(1 8 4 0 – 1 8 9 7)

From *Between the Flies and the Footlights*

DRUNKENNESS ON THE STAGE

It is always very difficult to represent drunkenness on the stage, the actor being drawn in different directions by the desire to be true to nature and the fear of offending good taste. For, in truth, how pitiable is the spectacle of that wilful debasement, of that temporary madness which man brings upon himself! To be sure, there is something comical in that self-abandonment of the human being, that faltering in speech and movement, in the awkward antics, the falls, the insane freaks of drunkenness, but the comicality is so heartrending that one can rarely disguise the distastefulness and horror of the situation with the aid of laughter.

On hearing Schneider, the illustrious diva of Meilhac and Halévy's operas,* stammer between two hiccoughs: *"I am a little tipsy; hush! you must not tell,"*—and seeing her fill the whole stage with her unsteady gait and her befuddled face, one could but think of the people coming out of a night restaurant in carnival time, when all the druggists' shops are closed, and they are unable, unfortunately, to procure a drop of ammonia.

*Hortense Schneider (1830–1920), soprano; Henri Meilhac (1831–1897), librettist who worked with Ludovic Halévy (1834–1908) on scores for Offenbach and Bizet, among which *Carmen* is their most famous.

On the other hand, how well Dupuis,* on the same stage, in *Les Millions de Gladiator,* acted the slight intoxication which follows a good dinner, at which one has drunk a little more than was necessary to quench his thirst; how eloquently young Isidore's tearful expansiveness, the mobility of his ideas, his tranquil contempt of life, bore witness to the generous and healthful qualities of the vintage he had abused!

And Bressant's drunkenness in the *Barber of Seville*—do you remember it?—what distinction, what good humor, what respect for truth and the proprieties!

Madame Marie-Laurent herself, before taking part in the *Voleuse d'Enfants,* had had a whole act of merry, extravagant, bumptious drunkenness in *Les Chevaliers du Brouillard;* but there she represented a young scamp embellished with all the vices, and the travesty facilitated the daring originality of the rôle. But to represent in Paris, before a French audience and at a time when comic opera had not made us proof against any eccentricity—to represent a woman too drunk to stand, was a difficult and ticklish undertaking. The actress hesitated a long while before undertaking to create the part; and when she had once made up her mind, she determined to cover up the odium of the impersonation by carrying it to that pitch of ghastly reality which becomes true art by virtue of accuracy, conscientiousness, and impulsive earnestness.

Her first idea was to go to London, to study the stupefying effects of gin in the slums of the great city; but as she had not the leisure for the journey, she contented herself with scrutinizing the common people of Paris, who, although they have no gin, have their vile barrier wines, pernicious and destructive, and absinthe and bitters, an endless variety of dangerous adulterations, which display their poisoned colors behind the dirty windows of the cabarets.

You should see the working-men on their way to work, at daybreak, on the outer boulevards, crowding around the doors of the wine-shops almost before they are open, and tossing off large glasses of white eau-de-vie—what they call "the drop"—to temper the cold, damp air of a Parisian morning. And such a drop! if a little of that

---

*Except for *The Barber of Seville* by Rossini, the performers and pieces remembered vividly by Daudet have been forgotten by history.

liquid overflows on the zinc counter, it leaves a corrosive blue stain, like the mark, still hot, made by a lighted match. Imagine that stuff pouring into a poor empty stomach. "It wakes you up!" as they say. Alas! it bestializes even more surely, and ere long the drunkenness of Paris will have no reason to envy the London article.

Often, on leaving the theatre, Marie-Laurent and her husband would follow for hours some wretched sot, who reeled against the wall as he walked, waving his hands, haranguing the closed doors, shouting his dream aloud, an incoherent dream, sometimes full of animation, sometimes melancholy. She would study the evolutions of that bewildered will, as it dragged the body in every direction, until at last, exhausted, vanquished, it propped him up against a post or stretched him at full length on the edge of the sidewalk, pale and dazed, with a fixed grimace of fatigue and suffering. Every day the artist observed some new detail, some new gesture, but as she departed from the conventional to enter into the real, she was more and more dismayed by the grim ghastliness of the task. "It isn't possible," she said to herself; "the audience will never let me go on to the end."

So it was that never, in any other of her creations, had she been assailed by such an overpowering dread as on the first night of the *Voleuse d'Enfants,* when she made her appearance in the sixth tableau. She entered at the rear, through a doorway several steps above the stage. Her alarm was heightened by the necessity of making that difficult descent characteristically, and according to rule.

Dressed in a marvellously hideous costume, all rags and tatters, horrible in her bewilderment and pallor, clinging to the rail, pitching forward, holding herself back, she reached the foot of the staircase without a sign from the audience to indicate its impression.

That glacial silence disturbed the actress. She had expected that as soon as she appeared the audience would be enthusiastic or disgusted, and would show it instantly.

Nothing of the sort. Utter stupefaction reigned supreme. The people watched and waited.

Oh, how long the descent of those six stairs seemed to her! "If I had walked from the Madeleine to the Bastille," she said afterward, "I should not have been so exhausted as when I reached the foot of that terrible staircase."

Those are, in very truth, terrible moments for the actor, who sees all those faces leaning forward or raised toward him, and those myriads of glances in which he can read naught save an expression of suspense, of eager but ill-defined curiosity.

But when she reached the front of the stage, when the audience, confronted by that ghastly image of drunkenness, by that pallid mask, distorted by horrible internal burns, those great eyes shooting flames, that black hair glued to the head by the mud of the gutter in which it had dragged again and again—when the audience suddenly realized that that bundle of rags was alive, aye, that it was suffering, and that they had not before them a vile sot but one of the damned, forgotten by God, who bore her hell within her, then they were deeply moved, they overflowed with pity and enthusiasm, and rewarded the brave actress with prolonged applause.

Y

CESARE PAVESE

(ITALIAN , 1 9 0 8 – 1 9 5 0)

The Old Drunk

Even the old woman likes stretching in the sun,
spreading out her arms. The heat is heavy,
pressing on her face as it presses on the ground.

Of things that warm, the only one left her is the sun.
Men and wine have betrayed, have consumed those brown
bones lying in the sun. But the cracked ground
hums like a flame. No words are needed,
no regrets. The quivering of a day returns
when her body too was young, hotter than the sun.

In her memories the huge hills swim into view,
young, alive, like that body of hers. And the man's look,
and the roughness of the wine, bring back the pain
of desire: a heat that quivered, humming in the blood
like greenness in the grass. Along paths and vineyards
her memory puts on flesh. Eyes shut, unmoving,
the old woman enjoys the sky with the body she used to have.

In the cracked earth beats a stronger heart
like the rugged chest of a husband or a father.
With her grizzled cheek she caresses it. In death
father and husband are both betrayed. Their flesh,
like hers, has been consumed. Neither warm thighs
nor rough wine will rouse them any more.

Far as the vineyards stretch, the sun's voice,
harsh and sweet, murmurs in the transparent blaze,
as if the air were quivering. The grass quivers around her.
The grass is as young as the heat of the sun.
The dead, in her green memories, grow young again.

—*Translated by William Arrowsmith*

Y

GUY DE MAUPASSANT

(1 8 5 0 – 1 8 9 3)

The Little Cask

Jules Chicot, the innkeeper, who lived at Épreville, pulled up his
tilbury in front of Mother Magloire's farmhouse. He was a tall
man of about forty, fat and with a red face and was generally said
to be a very knowing customer.

He hitched his horse up to the gate-post and went in. He owned some land adjoining that of the old woman. He had been coveting her plot for a long while, and had tried in vain to buy it a score of times, but she had always obstinately refused to part with it.

"I was born here, and here I mean to die," was all she said.

He found her peeling potatoes outside the farmhouse door. She was a woman of about seventy-two, very thin, shriveled and wrinkled, almost dried-up, in fact, and much bent, but as active and untiring as a girl. Chicot patted her on the back in a very friendly fashion, and then sat down by her on a stool.

"Well, Mother, you are always pretty well and hearty, I am glad to see."

"Nothing to complain of, considering, thank you. And how are you, Monsieur Chicot?"

"Oh! pretty well, thank you, except a few rheumatic pains occasionally; otherwise, I should have nothing to complain of."

"That's all the better!"

And she said no more, while Chicot watched her going on with her work. Her crooked, knotty fingers, hard as a lobster's claws, seized the tubers, which were lying in a pail, as if they had been a pair of pincers, and peeled them rapidly, cutting off long strips of skin with an old knife which she held in the other hand, throwing the potatoes into the water as they were done. Three daring fowls jumped one after another into her lap, seized a bit of peel and then ran away as fast as their legs would carry them with it in their beaks.

Chicot seemed embarrassed, anxious, with something on the tip of his tongue which he could not get out. At last he said hurriedly:

"I say, Mother Magloire—"

"Well, what is it?"

"You are quite sure that you do not want to sell your farm?"

"Certainly not; you may make up your mind to that. What I have said, I have said, so don't refer to it again."

"Very well; only I fancy I have thought of an arrangement that might suit us both very well."

"What is it?"

"Here you are: You shall sell it to me, and keep it all the same. You don't understand? Very well, so just follow me in what I am going to say."

The old woman left off peeling her potatoes and looked at the innkeeper attentively from under her bushy eyebrows, and went on:

"Let me explain myself: Every month I will give you a hundred and fifty francs.* You understand me, I suppose? Every month I will come and bring you thirty crowns,† and it will not make the slightest difference in your life—not the very slightest. You will have your own home just as you have now, will not trouble yourself about me, and will owe me nothing; all you will have to do will be to take my money: Will that arrangement suit you?"

He looked at her good-humoredly, one might have said benevolently, and the old woman returned his looks distrustfully, as if she suspected a trap, and said:

"It seems all right, as far as I am concerned, but it will not give you the farm."

"Never mind about that," he said, "you will remain here as long as it pleases God Almighty to let you live; it will be your home. Only you will sign a deed before a lawyer making it over to me after your death. You have no children, only nephews and nieces for whom you don't care a straw. Will that suit you? You will keep everything during your life, and I will give the thirty crowns a month. It is a pure gain as far as you are concerned."

The old woman was surprised, rather uneasy, but, nevertheless, very much tempted to agree and answered:

"I don't say that I will not agree to it, but I must think about it. Come back in a week and we will talk it over again, and I will then give you my definite answer."

And Chicot went off, as happy as a king who had conquered an empire.

Mother Magloire was thoughtful, and did not sleep at all that night; in fact, for four days she was in a fever of hesitation. She *smelled,* so to say, that there was something underneath the offer which was not to her advantage; but then the thought of thirty crowns a month, of all those coins chinking in her apron, falling to her, as it were, from the skies without her doing anything for it filled her with covetousness.

---

*About $30.
†The old name for a five-franc piece.

She went to the notary and told him about it. He advised her to accept Chicot's offer, but said she ought to ask for a monthly payment of fifty crowns instead of thirty, as her farm was worth sixty thousand francs at the lowest calculation.

"If you live fifteen years longer," he said, "even then he will only have paid forty-five thousand francs for it."

The old woman trembled with joy at this prospect of getting fifty crowns a month; but she was still suspicious, fearing some trick and she remained a long time with the lawyer asking questions without being able to make up her mind to go. At last she gave him instructions to draw up the deed, and returned home with her head in a whirl, just as if she had just drunk four jugs of new cider.

When Chicot came again to receive her answer she took a lot of persuading, and declared that she could not make up her mind to agree to his proposal, though she was all the time on tenterhooks lest he should not consent to give the fifty crowns. At last, when he grew urgent, she told him what she expected for her farm.

He looked surprised and disappointed, and refused.

Then, in order to convince him, she began to talk about the probable duration of her life.

"I am certainly not likely to live more than five or six years longer. I am nearly seventy-three, and far from strong, even considering my age. The other evening I thought I was going to die, and could hardly manage to crawl into bed."

But Chicot was not going to be taken in.

"Come, come, old lady, you are as strong as the church tower, and will live till you are a hundred at least; you will be sure to see me put underground first."

The whole day was spent in discussing the money, and as the old woman would not give way, the landlord consented to give the fifty crowns, and she insisted upon having ten crowns over and above to strike the bargain.

Three years passed by, and the old dame did not seem to have grown a day older. Chicot was in despair. It seemed to him as if he had been paying that annuity for fifty years, that he had been taken in, outwitted, and ruined. From time to time he went to see his annuitant, just as one goes in July to see when the harvest is likely to begin. She always met him with a cunning look, and one would have felt inclined to think that she was congratulating herself on the trick she had played on him. Seeing how well and hearty she seemed, he very soon got into his tilbury again, growling to himself:

"Will you never die, you old brute?"

He did not know what to do, and felt inclined to strangle her when he saw her. He hated her with a ferocious, cunning hatred, the hatred of a peasant who has been robbed, and began to cast about for means of getting rid of her.

One day he came to see her again, rubbing his hands like he did the first time when he proposed the bargain, and, after having chatted for a few minutes, he said:

"Why do you never come and have a bit of dinner at my place when you are in Épreville? The people are talking about it and saying that we are not on friendly terms, and that pains me. You know it will cost you nothing if you come, for I don't look at the price of a dinner. Come whenever you feel inclined; I shall be very glad to see you."

Old Mother Magloire did not need to be told twice, and the next day but one—she was going to the town in any case, it being market-day, in her gig, driven by her man—she, without any demur, put her trap up in Chicot's stable, and went in search of her promised dinner.

The publican was delighted, and treated her like a princess, giving her roast fowl, black pudding, leg of mutton, and bacon and cabbage. But she ate next to nothing. She had always been a small eater and had generally lived on a little soup and a crust of bread-and-butter.

Chicot was disappointed, and pressed her to eat more, but she refused. She would drink next to nothing either, and declined any coffee, so he asked her:

"But surely, you will take a little drop of brandy or liquor?"

"Well, as to that, I don't know that I will refuse." Whereupon he shouted out:

"Rosalie, bring the superfine brandy,—*the special*—you know."

The servant appeared, carrying a long bottle ornamented with a paper vine-leaf, and he filled two liquor glasses.

"Just try that; you will find it first-rate."

The good woman drank it slowly in sips, so as to make the pleasure last all the longer, and when she had finished her glass, draining the last drops so as to make sure of all, she said:

"Yes, that is first-rate!"

Almost before she had said it, Chicot had poured her out another glassful. She wished to refuse, but it was too late, and she drank it very slowly, as she had done the first, and he asked her to have a third. She objected, but he persisted.

"It is as mild as milk, you know. I can drink ten or a dozen without any ill effect; it goes down like sugar, and leaves no headache behind; one would think that it evaporated on the tongue. It is the most wholesome thing you can drink."

She took it, for she really wished to have it, but she left half the glass.

Then Chicot, in an excess of generosity said:

"Look here, as it is so much to your taste, I will give you a small keg of it, just to show that you and I are still excellent friends." Then she took her leave, feeling slightly overcome by the effects of what she had drunk.

The next day the innkeeper drove into her yard, and took a little iron-hooped keg out of his gig. He insisted on her tasting the contents, to make sure it was the same delicious article, and, when they had each of them drunk three more glasses, he said, as he was going away:

"Well, you know, when it is all gone, there is more left; don't be modest for I shall not mind. The sooner it is finished the better pleased I shall be."

Four days later he came again. The old woman was outside her door cutting up the bread for her soup.

He went up to her, and put his face close to hers, so that he might smell her breath; and when he smelled the alcohol he felt pleased.

"I suppose you will give me a glass of *the special?*" he said. And they had three glasses each.

Soon, however, it began to be whispered abroad that Mother

Magloire was in the habit of getting drunk all by herself. She was picked up in her kitchen, then in her yard, then in the roads in the neighborhood, and was often brought home like a log.

Chicot did not go near her any more, and when people spoke to him about her, he used to say, putting on a distressed look:

"It is a real pity that she should have taken to drink at her age; but when people get old there is no remedy. It will be the death of her in the long run."

And it certainly was the death of her. She died the next winter. About Christmas time she fell down unconscious in the snow, and was found dead the next morning.

And when Chicot came in for the farm he said:

"It was very stupid of her; if she had not taken to drink she might very well have lived for ten years longer."

—*Translated by Walter Dunn*

Y

F. SCOTT FITZGERALD

(AMERICAN, 1896–1940)

An Alcoholic Case

I

"Let—go—that—oh-h-h! Please, now, will you? *Don't* start drinking again! Come on—give me the bottle. I told you I'd stay awake givin it to you. Come on. If you do like that a-way—then what are you going to be like when you go home. Come on—leave it with me—I'll leave half in the bottle. Pul-lease. You know what Dr. Carter says—I'll stay awake and give it to you, or else fix some of it in the bottle—come on—like I told you, I'm too tired to be fightin you all night. . . . All right, drink your fool self to death."

"Would you like some beer?" he asked.

"No, I don't want any beer. Oh, to think that I have to look at you drunk again. My God!"

"Then I'll drink the Coca-Cola."

The girl sat down panting on the bed.

"Don't you believe in anything?" she demanded.

"Nothing you believe in—please—it'll spill."

She had no business there, she thought, no business trying to help him. Again they struggled, but after this time he sat with his head in his hands awhile, before he turned around once more.

"Once more you try to get it I'll throw it down," she said quickly. "I will—on the tiles in the bathroom."

"Then I'll step on the broken glass—or you'll step on it."

"Then let go—oh you promised—"

Suddenly she dropped it like a torpedo, sliding underneath her hand and slithering with a flash of red and black and the words: SIR GALAHAD, DISTILLED LOUISVILLE GIN. He took it by the neck and tossed it through the open door to the bathroom.

It was on the floor in pieces and everything was silent for awhile and she read *Gone With the Wind* about things so lovely that had happened long ago. She began to worry that he would have to go into the bathroom and might cut his feet, and looked up from time to time to see if he would go in. She was very sleepy—the last time she looked up he was crying and he looked like an old Jewish man she had nursed once in California; he had had to go to the bathroom many times. On this case she was unhappy all the time but she thought:

"I guess if I hadn't liked him I wouldn't have stayed on the case."

With a sudden resurgence of conscience she got up and put a chair in front of the bathroom door. She had wanted to sleep because he had got her up early that morning to get a paper with the story of the Yale-Dartmouth game in it and she hadn't been home all day. That afternoon a relative of his had come to see him and she had waited outside in the hall where there was a draft with no sweater to put over her uniform.

As well as she could she arranged him for sleeping, put a robe over his shoulders as he sat slumped over his writing table, and one on his knees. She sat down in the rocker but she was no longer sleepy; there was plenty to enter on the chart and treading lightly about she found a pencil and put it down:

Pulse 120
Respiration 25
Temp. 98—98.4—98.2
Remarks—
—She could make so many:
Tried to get bottle of gin. Threw it away and broke it.
She corrected it to read:
In the struggle it dropped and was broken. Patient was generally difficult.

She started to add as part of her report: *I never want to go on an alcoholic case again,* but that wasn't in the picture. She knew she could wake herself at seven and clean up everything before his niece awakened. It was all part of the game. But when she sat down in the chair she looked at his face, white and exhausted, and counted his breathing again, wondering why it had all happened. He had been so nice today, drawn her a whole strip of his cartoon just for fun and given it to her. She was going to have it framed and hang it in her room. She felt again his thin wrists wrestling against her wrist and remembered the awful things he had said, and she thought too of what the doctor had said to him yesterday: "You're too good a man to do this to yourself."

She was tired and didn't want to clean up the glass on the bathroom floor, because as soon as he breathed evenly she wanted to get him over to the bed. But she decided finally to clean up the glass first; on her knees, searching a last piece of it, she thought:

—This isn't what I ought to be doing. And this isn't what *he* ought to be doing.

Resentfully she stood up and regarded him. Through the thin delicate profile of his nose came a light snore, sighing, remote, inconsolable. The doctor had shaken his head in a certain way, and she knew that really it was a case that was beyond her. Besides, on her card at the agency was written, on the advice of her elders, "No Alcoholics."

She had done her whole duty, but all she could think of was that when she was struggling about the room with him with that gin bottle there had been a pause when he asked her if she had hurt her elbow against a door and that she had answered: "You don't know how people talk about you, no matter how you think of yourself—" when she knew he had a long time ceased to care about such things.

The glass was all collected—as she got out a broom to make sure, she realized that the glass, in its fragments, was less than a window through which they had seen each other for a moment. He did not know about her sisters, and Bill Markoe whom she had almost married, and she did not know what had brought him to this pitch, when there was a picture on his bureau of his young wife and his two sons and him, all trim and handsome as he must have been five years ago. It was so utterly senseless—as she put a bandage on her finger where she had cut it while picking up the glass she made up her mind she would never take an alcoholic case again.

II

It was early the next evening. Some Halloween jokester had split the side windows of the bus and she shifted back to the Negro section in the rear for fear the glass might fall out. She had her patient's check but no way to cash it at this hour; there was a quarter and a penny in her purse.

Two nurses she knew were waiting in the hall of Mrs. Hixson's Agency.

"What kind of case have you been on?"

"Alcoholic," she said.

"Oh yes—Gretta Hawks told me about it—you were on with that cartoonist who lives at the Forest Park Inn."

"Yes, I was."

"I hear he's pretty fresh."

"He's never done anything to bother me," she lied. "You can't treat them as if they were committed—"

"Oh, don't get bothered—I just heard that around town—oh, you know—they want you to play around with them—"

"Oh, be quiet," she said, surprised at her own rising resentment.

In a moment Mrs. Hixson came out and, asking the other two to wait, signaled her into the office.

"I don't like to put young girls on such cases," she began. "I got your call from the hotel."

"Oh, it wasn't bad, Mrs. Hixson. He didn't know what he was

doing and he didn't hurt me in any way. I was thinking much more of my reputation with you. He was really nice all day yesterday. He drew me—"

"I didn't want to send you on that case." Mrs. Hixson thumbed through the registration cards. "You take T.B. cases, don't you? Yes, I see you do. Now here's one—"

The phone rang in a continuous chime. The nurse listened as Mrs. Hixson's voice said precisely:

"I will do what I can—that is simply up to the doctor. . . . That is beyond my jurisdiction. . . . Oh, hello, Hattie, no, I can't now. Look, have you got any nurse that's good with alcoholics? There's somebody up at the Forest Park Inn who needs somebody. Call back will you?"

She put down the receiver. "Suppose you wait outside. What sort of man is this, anyhow? Did he act indecently?"

"He held my hand away," she said, "so I couldn't give him an injection."

"Oh, an invalid he-man," Mrs. Hixson grumbled. "They belong in sanitaria. I've got a case coming along in two minutes that you can get a little rest on. It's an old woman—"

The phone rang again. "Oh, hello, Hattie. . . . Well, how about that big Svensen girl? She ought to be able to take care of any alcoholic. . . . How about Josephine Markham? Doesn't she live in your apartment house? . . . Get her to the phone." Then after a moment, "Joe, would you care to take the case of a well-known cartoonist, or artist, whatever they call themselves, at Forest Park Inn? . . . No, I don't know, but Dr. Carter is in charge and will be around about ten o'clock."

There was a long pause; from time to time Mrs. Hixson spoke:

"I see. . . . Of course, I understand your point of view. Yes, but this isn't supposed to be dangerous—just a little difficult. I never like to send girls to a hotel because I know what riff-raff you're liable to run into. . . . No, I'll find somebody. Even at this hour. Never mind and thanks. Tell Hattie I hope the hat matches the negligee. . . ."

Mrs. Hixson hung up the receiver and made notations on the pad before her. She was a very efficient woman. She had been a nurse and had gone through the worst of it, had been a proud,

idealistic, overworked probationer, suffered the abuse of smart in-
ternes and the insolence of her first patients, who thought that she
was something to be taken into camp immediately for premature
commitment to the service of old age. She swung around suddenly
from the desk.

"What kind of cases do you want? I told you I have a nice old
woman—"

The nurse's brown eyes were alight with a mixture of thoughts—
the movie she had just seen about Pasteur and the book they had
all read about Florence Nightingale when they were student nurses.
And their pride, swinging across the streets in the cold weather at
Philadelphia General, as proud of their new capes as debutantes in
their furs going in to balls at the hotels.

"I—I think I would like to try the case again," she said amid a
cacophony of telephone bells. "I'd just as soon go back if you can't
find anybody else."

"But one minute you say you'll never go on an alcoholic case
again and the next minute you say you want to go back to one."

"I think I overestimated how difficult it was. Really, I think I
could help him."

"That's up to you. But if he tried to grab your wrists."

"But he couldn't," the nurse said. "Look at my wrists: I played
basketball at Waynesboro High for two years. I'm quite able to
take care of him."

Mrs. Hixson looked at her for a long minute. "Well, all right,"
she said. "But just remember that nothing they say when they're
drunk is what they mean when they're sober—I've been all through
that; arrange with one of the servants that you can call on him,
because you never can tell—some alcoholics are pleasant and some
of them are not, but all of them can be rotten."

"I'll remember," the nurse said.

It was an oddly clear night when she went out, with slanting
particles of thin sleet making white of a blue-black sky. The bus
was the same that had taken her into town, but there seemed to
be more windows broken now and the bus driver was irritated
and talked about what terrible things he would do if he caught
any kids. She knew he was just talking about the annoyance in
general, just as she had been thinking about the annoyance of an

alcoholic. When she came up to the suite and found him all helpless and distraught she would despise him and be sorry for him.

Getting off the bus, she went down the long steps to the hotel,
feeling a little exalted by the chill in the air. She was going to take
care of him because nobody else would, and because the best people of her profession had been interested in taking care of the cases
that nobody else wanted.

She knocked at his study door, knowing just what she was going
to say.

He answered it himself. He was in dinner clothes even to a
derby hat—but minus his studs and tie.

"Oh, hello," he said casually. "Glad you're back. I woke up a
while ago and decided I'd go out. Did you get a night nurse?"

"I'm the night nurse too," she said. "I decided to stay on twenty-
four hour duty."

He broke into a genial, indifferent smile.

"I saw you were gone, but something told me you'd come back.
Please find my studs. They ought to be either in a little tortoise
shell box or—"

He shook himself a little more into his clothes, and hoisted the
cuffs up inside his coat sleeves.

"I thought you had quit me," he said casually.

"I thought I had, too."

"If you look on that table," he said, "you'll find a whole strip of
cartoons that I drew you."

"Who are you going to see?" she asked.

"It's the President's secretary," he said. "I had an awful time
trying to get ready. I was about to give up when you came in.
Will you order me some sherry?"

"One glass," she agreed wearily.

From the bathroom he called presently:

"Oh, nurse, nurse, Light of my Life, where is another stud?"

"I'll put it in."

In the bathroom she saw the pallor and the fever on his face
and smelled the mixed peppermint and gin on his breath.

"You'll come up soon?" she asked. "Dr. Carter's coming at ten."

"What nonsense! You're coming down with me."

"Me?" she exclaimed. "In a sweater and skirt? Imagine!"

"Then I won't go."

"All right then, go to bed. That's where you belong anyhow. Can't you see these people tomorrow?"

"No, of course not."

"Of course not!"

She went behind him and reaching over his shoulder tied his tie—his shirt was already thumbed out of press where he had put in the studs, and she suggested:

"Won't you put on another one, if you've got to meet some people you like?"

"All right, but I want to do it myself."

"Why can't you let me help you?" she demanded in exasperation. "Why can't you let me help you with your clothes? What's a nurse for—what good am I doing?"

He sat down suddenly on the toilet seat.

"All right—go on."

"Now don't grab my wrist," she said, and then, "Excuse me."

"Don't worry. It didn't hurt. You'll see in a minute."

She had the coat, vest and stiff shirt off him but before she could pull his undershirt over his head he dragged at his cigarette, delaying her.

"Now watch this," he said. "One—two—three."

She pulled up the undershirt; simultaneously he thrust the crimson-gray point of the cigarette like a dagger against his heart. It crushed out against a copper plate on his left rib about the size of a silver dollar, and he said "ouch!" as a stray spark fluttered down against his stomach.

Now was the time to be hardboiled, she thought. She knew there were three medals from the war in his jewel box, but she had risked many things herself: tuberculosis among them and one time something worse, though she had not known it and had never quite forgiven the doctor for not telling her.

"You've had a hard time with that, I guess," she said lightly as she sponged him. "Won't it ever heal?"

"Never. That's a copper plate."

"Well, it's no excuse for what you're doing to yourself."

He bent his great brown eyes on her, shrewd—aloof, confused. He signaled to her, in one second, his Will to Die, and for all her training and experience she knew she could never do anything constructive with him. He stood up, steadying himself on the washbasin and fixing his eye on some place just ahead.

"Now, if I'm going to stay here you're not going to get at that liquor," she said.

Suddenly she knew he wasn't looking for that. He was looking at the corner where he had thrown the bottle the night before. She stared at his handsome face, weak and defiant—afraid to turn even halfway because she knew that death was in that corner where he was looking. She knew death—she had heard it, smelt its unmistakable odor, but she had never seen it before it entered into anyone, and she knew this man saw it in the corner of his bathroom; that it was standing there looking at him while he spit from a feeble cough and rubbed the result into the braid of his trousers. It shone there . . . crackling for a moment as evidence of the last gesture he ever made.

She tried to express it next day to Mrs. Hixson:

"It's not like anything you can beat—no matter how hard you try. This one could have twisted my wrists until he strained them and that wouldn't matter so much to me. It's just that you can't really help them and it's so discouraging—it's all for nothing."

HENRY LAWSON

(AUSTRALIAN, 1867–1922)

The Boozers' Home

"A dipsomaniac," said Mitchell, "needs sympathy and commonsense treatment. (Sympathy's a grand and glorious thing, taking it all round and looking at it any way you will: a little of it makes a man think that the world's a good world after all, and there's room and hope for sinners, and that life's worth living; enough of it makes him sure of it: and an overdose of sympathy makes a man *feel* weak and ashamed of himself, and so moves him to stop whining—and wining—and buck up.)

"Now, I'm not taking the case of a workman who goes on the spree on pay night and sweats the drink out of himself at work next day, nor a slum-bred brute who guzzles for the love of it; but a man with brains, who drinks to drown his intellect or his memory. He's generally a man under it all, and a sensitive, generous, gentle man with finer feelings as often as not. The best and cleverest and whitest men in the world seem to take to drink mostly. It's an awful pity. Perhaps it's because they're straight and the world's crooked and they can see things too plain. And I suppose in the Bush the loneliness and the thoughts of the girl-world they left behind help to sink 'em.

"Now a drunkard seldom reforms at home, because he's always surrounded by the signs of the ruin and misery he has brought on the home; and the sight and thought of it sets him off again before he's had time to recover from the last spree. Then, again, the noblest wife in the world mostly goes the wrong way to work with a drunken husband—nearly everything she does is calculated to irritate him. If, for instance, he brings a bottle home from the pub, it shows that he wants to stay at home and not go back to the pub any more; but the first thing the wife does is to get hold of the bottle and plant it, or smash it before his eyes, and that maddens him in the state he is in then.

"No. A dipsomaniac needs to be taken away from home for a while. I knew a man that got so bad that the way he acted at home one night frightened him, and next morning he went into an inebriate home of his own accord—to a place where his friends had been trying to get him for a year past. For the first day or two he was nearly dead with remorse and shame—mostly shame; and he didn't know what they were going to do to him next—he only wanted them to kill him quick and be done with it. He reckons he felt as bad as if he was in jail. But there were ten other patients there, and one or two were worse than he was, and that comforted him a lot. They compared notes and sympathised and helped each other. They discovered that all their wives were noble women. He struck one or two surprises too—one of the patients was a doctor who'd attended him one time, and another was an old boss of his, and they got very chummy. And there was a man there who was standing for Parliament—he was supposed to be having a rest down the coast. . . . Yes, my old mate felt very bad for the first day or two; it was all Yes, Nurse, and Thank you, Nurse, and Yes, Doctor, and No, Doctor, and Thank you, Doctor. But, inside a week, he was calling the doctor 'Ol' Pill-Box' behind his back, and making love to one of the nurses.

"But he said it was pitiful when women relatives came to visit patients the first morning. It shook the patients up a lot, but I reckon it did 'em good. There were well-bred old lady mothers in black, and hard-working, haggard wives and loving daughters— and the expressions of sympathy and faith and hope in those women's faces! My old mate said it was enough in itself to make a man swear off drink forever. . . . Ah, God—what a world it is!

"Reminds me how I once went with the wife of another old mate of mine to see him. He was in a lunatic asylum. It was about the worst hour I ever had in my life, and I've had some bad ones. The way she tried to coax him back to his old self. She thought she could do it when all the doctors had failed. But I'll tell you about him some other time.

"The old mate said that the principal part of the treatment was supposed to be injection of bi-chloride of gold or something, and it was supposed to be a secret. It might have been water and sugar for all he knew, and he thought it was. You see, when patients got better they were allowed out, two by two, on their honour— one to watch the other—and it worked. But it was necessary to have an extra hold on them; so they were told that if they were a minute late for 'treatment,' or missed one injection, all the good would be undone. This was dinged into their ears all the time. Same as many things are done in the Catholic religion—to hold the people. My old mate said that, as far as the medical treatment was concerned, he could do all that was necessary himself. But it was the sympathy that counted, especially the sympathy between the patients themselves. They always got hold of a new patient and talked to him and cheered him up; he nearly always came in thinking he was the most miserable wretch in the world. And it comforts a man and strengthens him and makes him happier to meet another man who's worse off or sicker, or has been worse swindled than he has been. That's human nature. . . . And a man will take draughts from a nurse and eat for her when he wouldn't do it for his own wife—not even though she had been a trained nurse herself. And if a patient took a bad turn in the night at the Boozers' Home and got up to hunt the snakes out of his room, he wouldn't be sworn at, or laughed at, or held down; no, they'd help him shoo the snakes out and comfort him. My old mate said that, when he got better, one of the new patients reckoned that he licked St. Pathrick at managing snakes. And when he came out he didn't feel a bit ashamed of his experience. The institution didn't profess to cure anyone of drink, only to mend up shattered nerves and build up wrecked constitutions; give them back some will power if they weren't too far gone. And they set my old mate on his feet all right. When he went in his life seemed lost, he had the horror of being sober, he couldn't start the day without a drink or do any business without it. He couldn't live for more than two hours without a drink; but when he came out he didn't feel as if he wanted it. He reckoned that those six weeks in the institution were the happiest he'd ever spent in his life, and he wished the time had been longer; he says he'd never met with so much sympathy and genius, and humour and human nature, under one roof before. And he said it was nice and novel to be looked after and watched and physicked and bossed by a pretty nurse in uniform—but I don't suppose he told his wife that. And when he came out he never took the trouble to hide the fact that he'd been in. If any of his friends had a drunkard in the family, he'd recommend the institution and do his best to get him into it. But when he came out he firmly believed that if he took one drink he'd be a lost man. He made a mania of that. One curious effect was that, for some time after he left the institution, he'd sometimes feel suddenly in high spirits—with nothing to account for it—something like he used to feel when he had half a dozen whiskies in him; then suddenly he'd feel depressed and sort of hopeless—with nothing to account for that either—just as if he was suffering a recovery. But those moods never lasted long and he soon grew out of them altogether. He didn't flee temptation. He'd knock round the pubs on Saturday nights with his old mates, but never drank anything but soft stuff—he was always careful to smell his glass for fear of an accident or a trick. He drank gallons of ginger-beer, milk-and-soda and lemonade; and he got very fond of sweets, too—he'd never liked them before. He said he enjoyed the novelty of the whole thing and his mates amused him at first; but he found he had to leave them early in the evening, and, after a while, he dropped them altogether. They seemed such fools when they were drunk (they'd never seemed fools to him before). And, besides, as they got full, they'd get suspicious of him, and then mad at him, because he couldn't see things as they could. That reminds me that it nearly breaks a man's heart when his old drinking chum turns teetotaller—it's worse than if he got married or died. When two mates meet and one is drunk and the other sober there is only one of two things for them to do if they want to hit it together—either the drunken mate must get sober or the sober mate drunk. And that reminds me: Take the case of two old mates who've been together all their lives, say they always had their regular sprees together and went through the same stages of drunkenness together, and suffered their recoveries and sobered up together, and each could stand about the same quantity of drink and one never got drunker than the other. Each, when he's boozing, reckons his mate the cleverest man and the hardest case in the world—second to himself. But one day it happens, by a most extraordinary combination of circumstances, that Bill, being sober, meets Jim very drunk, and pretty soon Bill is the most disgusted man in this world. He never would have dreamed that his old mate could make such a fool and such a public spectacle of himself. And Bill's disgust intensifies all the time he is helping Jim home, and Jim arguing with him and wanting to fight him, and slobbering over him and wanting to love him by turns, until Bill swears he'll give Jim a hammering as soon as ever he's able to stand steady on his feet."

"I suppose your old boozing mate's wife was very happy when he reformed," I said to Mitchell.

"Well, no," said Mitchell, rubbing his head rather ruefully. "I suppose it was an exceptional case. But I knew her well, and the fact is that she got more discontented and thinner, and complained and nagged him worse than she'd ever done in his drinking days. And she'd never been afraid of him. Perhaps it was this way: She loved and married a careless, good-natured, drinking scamp, and when he reformed and became a careful, hard-working man, and an honest and respected fellow-townsman, she was disappointed in him. He wasn't the man that won her heart when she was a girl. Or maybe he was only company for her when he was half drunk. Or maybe lots of things. Perhaps he'd killed the love in her before he reformed—and reformed too late. I wonder how a man feels when he finds out for the first time that his wife doesn't love him any longer? But my old mate wasn't the nature to find out that sort of thing. Ah, well! If a woman caused all our trouble, my God! women have suffered for it since—and they suffer like martyrs mostly and with the patience of working bullocks. Anyway it goes, if I'm the last man in the world, and the last woman is the worst, and there's only room for one more in Heaven, I'll step down at once and take my chance in Blazes."

VACHEL LINDSAY

(AMERICAN, 1879–1931)

The Drunkard's Funeral

"Yes," said the sister with the little pinched face,
The busy little sister with the funny little tract:—
"This is the climax, the grand fifth act.
There rides the proud, at the finish of his race.
There goes the hearse, the mourners cry,
The respectable hearse goes slowly by.
The wife of the dead has money in her purse,
The children are in health, so it might have been worse.
The fellow in the coffin led a life most foul.
A fierce defender of the red bartender,
At the church he would rail,
At the preacher he would howl.

He planted every deviltry to see it grow.
He wasted half his income on the lewd and the low.
He would trade engender for the red bar-tender,
He would homage render to the red bar-tender,
And in ultimate surrender to the red bar-tender,
He died of the tremens, as crazy as a loon,
And his friends were glad, when the end came soon.
There goes the hearse, the mourners cry,
The respectable hearse goes slowly by.
And now, good friends, since you see how it ends,
Let each nation-mender flay the red bar-tender—
Abhor
The transgression
Of the red bar-tender—
Ruin
The profession
Of the red bar-tender:
Force him into business where his work does good.
Let him learn how to plough, let him learn to chop wood,
Let him learn how to plough, let him learn to chop wood.

"The moral,
The conclusion,
The verdict now you know:—
'The saloon must go,
The saloon must go,
The saloon,
The saloon,
The saloon,
Must go.'"
"You are right, little sister," I said to myself,
"You are right, good sister," I said.
"Though you wear a mussy bonnet
On your little gray head,
You are right, little sister," I said.

Y

GEORGE ADE

(AMERICAN, 1866–1944)

R-e-m-o-r-s-e

The cocktail is a pleasant drink,
It's mild and harmless, I don't think.
When you've had one, you call for two,
And then you don't care what you do.
Last night I hoisted twenty-three
Of these arrangements into me;
My wealth increased, I swelled with pride;
I was pickled, primed and ossified.

R-E-M-O-R-S-E!
Those dry martinis did the work for me;
Last night at twelve I felt immense;
Today I feel like thirty cents.
At four I sought my whirling bed,
At eight I woke with such a head!
It is no time for mirth or laughter—
The cold, gray dawn of the morning after.

If ever I want to sign the pledge,
It's the morning after I've had an edge;
When I've been full of the oil of joy
And fancied I was a sporty boy.
This world was one kaleidoscope
Of purple bliss, transcendent hope.
But now I'm feeling mighty blue—
Three cheers for the W.C.T.U.!*

*Women's Christian Temperance Union.

R-E-M-O-R-S-E!
The water wagon is the place for me;
I think that somewhere in the game,
I wept and told my maiden name.
My eyes are bleared, my coppers hot;
I try to eat, but I can not;
It is no time for mirth or laughter—
The cold, gray dawn of the morning after.

AMERICAN POPULAR SONG

The Tee-to-tal* Society

I'm come to exhort you so free, all you that so fond of the bottle
    are,
And when you my argument see, ev'ry one will become a Teeto-
    taler
Of gin, brandy, rum, wine, or beer to drink is a great impro-
    priety,
Of trash I'd have you steer clear, and join the *Teetotal Society*.

An old man that was troubled with corns, that scarcely the stairs
    could he hobble up,
He used to drink beer out of horns, and all sorts of liquors
    would gobble up.
His corns have all left one by one, and now he's the pink of so-
    briety.
And pray how was all this done? *Why?* he joined the *Teetotal So-
    ciety*.

* That is, abstinent to the "T," or totally!

Tother day my young pigs and old sow, I found to be far gone
   in liquor,
In my family I this can't allow, to temperance being a sticker,
They had with grain from the brewer been fed, but now they
   shall share in sobriety,
Coffee grounds and tea leaves instead, and they shall join the
   Teetotal Society.

A Teetotaler tother day died, the doctor his friend did entice out
Examined his stomach inside, and they say took a large lump of
   ice out.
This cant be true, for if ever we are ill, of Brandy we take a small
   moiety
And melt ice you know, *Brandy* it will, it has been tried by the
   Teetotal Society.

I wander about doing good, our society pays all my charges
Preaching two hours at least, to coal heavers working on barges.
But they said, "If you carried our coals, of beer you'll soon see
   the propriety."
But ah! they are sad wicked souls, they wont join the *Teetotal
   Society.*

Folks ask what makes my nose so red? I'll tell and end all this
   puzzling,
It'ant drink what gets in my head, its blushing to see so much
   guzzling.
Drops of Brandy we take two or three, as medicine and no impro-
   priety
And put some in our gruel and tea, its allowed by the *Teetotal
   Society.*

The people laugh at me oh! dear, and puts my mind in sad order
   works
And cries out whenever I appear, "How gets on the Temperance
   water works."
But I tells them I dont care dump, and preaches away on sobriety
And for example drinks out of the pump, since I joined the *Teetotal Society.*

In our progress there's nothing excells, in our efforts we never do
    slumber, sir,
We have dug six and fifty new wells! and erected of pumps a
    great number, sir,
I have here some Temperance tracts, of a most gratifying variety
That record some most wonder facts, about the *Teetotal Society*.

A drunken beggar I very well know, quite lame and as thin as a
    rat he was
Led by a dog he would go, through the streets, for blind as a bat
    he was.
You'll scarcely believe what I say, he's now the pink of sobriety
He's got fat and can see as clear as day, since he joined the *Teeto-
tal Society*.

One night in my house every week, I holds forth on the beauties
    of Temperance
Because when in public I speak, I'm subject to a good deal of
    imperance.
After a lecture on Coffee they sup, on Tea if they like for variety.
I charge a shilling a cup, since I joined the *Teetotal Society*.

❦

DON MARQUIS

(AMERICAN, 1878–1937)

Mrs. Swartz

"WHEELING, W. VA.—Prohibition agents who surprised Mrs. Mike Swartz, of Boggs Run, while she was operating a still to-day, heeded the pleas of her nine children and refused to arrest her. Mike Swartz, her husband, is in a Federal prison for robbing a freight car."—*News story.*

1

The Prohibition agents came
Unto a cabin door,
Nine angel children played their games
And romped upon the floor;
The agents laid a burly hand
On Mother's hair so gray,
For making hootch,* you understand,
And all the Tots did say:

Chorus:
"Oh, do not take our Mother's still,
for she is old and worn,
What will she do if she can't make
the moonshine from the corn†
Oh, do not lock our Mother up!
What will become of we
Without the hootch we learned to drink
at dear old Mother's knee?"

2

The Prohibition agents then
Felt tears fall down their cheek,
At heart they were not wicked men,
Oh, they were only weak!
And both removed their burly hands
From Mother's hair so gray;
They felt remorse, you understand,
And to the Tots did say:

Chorus:
"We will not take your Mother's still,
for she is old and worn!
What will she do if she can't make
the moonshine from the corn?

* Illegal whiskey, named after the Alaskan Hoochinoo, who distilled spirits illegally.
† She is making bourbon, or "sour mash" whiskey.

We will not lock your Mother up.
  What would become of thee
Without the hootch thou learned to
  drink at dear old Mother's knee?"

3

The eldest daughter thanked them then,
  A maid of sweet sixteen,
Who had not often spoke to men,
  She was a little queen.
She smiled at them above her tears,
  So fairylike and gay,
And said, "I'll thank you all my years,
  Because you just did say:

Chorus:
  "We will not take your Mother's still," etc.

4

The youngest agent spoke to her:
  "You look so sweet and pure!
What man would be so low a cur
  As think you to injure?
I offer you on bended knee
  All of my manly love!"
An Angel Voice that sang in glee
  Came to them from Above:

Chorus:
  "They will not take the Mother's still,"

5

Just then a lusty shout was heard
  Outside the cabin door,
A clean-shaved man with coat all furred
  Came riding in a Ford.
It was their Father, who'd escaped
  From jail that very day,

And kneeling down among them prayed
While all of them did say:

Chorus:
"They have not took the Mother's still,
for she is old and worn!
What could she do if she don't make
the moonshine from the corn?
They have not locked the Mother up!
What would become of thee
Without the hootch thou learned to
drink
At dear old Mother's knee?"

♆

JAROSLAV HAŠEK

(C Z E C H , 1 8 8 3 – 1 9 2 3)

From *The Good Soldier Švejk*

If you had known old Vejvoda, sir, a builder's foreman from Vršovice*—he once took it into his head that he wouldn't drink any drink which would make him drunk. And so he had his last wee dram for the road at home and set out on a journey to find some non-alcoholic drinks. He called first of all at a pub called At the Stop, had a quarter-litre of vermouth there and started his unobtrusive inquiries of the landlord about what sort of stuff those total abstainers actually drank. He was quite right in thinking that pure water was rather cruel fare even for total abstainers. The landlord then explained to him that total abstainers drink soda

* Southeast suburb of Prague, Czechoslovakia.

water, mineral water, milk and then various kinds of alcohol-free wines, cold clear soup and other beverages without spirit. Of this wide range of drinks it was the alcohol-free wine which most appealed to Vejvoda's taste. He asked one more question: whether there also existed spirits without alcohol. Then he had another quarter-litre of vermouth, talked with the landlord about how it was a real sin to get tight very often, whereupon the latter told him that he could endure everything in the world except a chap who goes and gets himself sozzled in another pub and only comes to him to get sober with a bottle of soda water and perhaps makes a terrible row as well. "Get sozzled in my pub," said the landlord, "and you are my man, but otherwise I won't have anything to do with you." Old Vejvoda finished his vermouth and went out to continue his journey until he came—just imagine it, sir—to a wine shop at Charles Square which he also used to visit from time to time. He asked there whether they didn't have wines without spirit. "Sorry, Mr. Vejvoda," they told him, "we don't have any wines without spirit, just vermouth or sherry." Somehow or other old Vejvoda felt ashamed and so he had a quarter-litre of vermouth and a quarter-litre of sherry, and as he sat there he met one of those total abstainers. They got talking and drank another quarter-litre of sherry, and in the end it turned out that the gentleman knew a place where they served wines without spirit. "It's in Bolzanova Street," he said. "You get there down the steps and they have a gramophone there." In reward for this cheering information old Vejvoda ordered a whole bottle of vermouth and then they both walked to the place in Bolzanova Street, which was down the steps and where they had a gramophone. And it was quite true that they only served fruit wines there, not only free of spirit but without alcohol either. First of all each ordered a half-litre of gooseberry wine and then a half-litre of redcurrant wine, and when they had drunk another half-litre of spirit-free gooseberry wine they began to feel pins and needles in their legs after all the vermouth and sherry they had had before. And they began to shout that they must be given official confirmation that what they were drinking there was spirit-free wine. They were total abstainers and if they didn't get this at once they would break everything up including the gramophone. Finally the cops had to drag both of them up the steps again to Bolzanova Street. And they had to put them in the drunks' cart and chuck them into the isolation cell. Both had to be sentenced for being drunk and disorderly when they were total abstainers.

—*Translated by Cecil Parrott*

Y

O. HENRY

(A M E R I C A N , 1 8 6 2 – 1 9 1 0)

The Pint Flask

A prominent Houston colonel, who is also a leading church member, started for church last Sunday morning with his family, as was his custom. He was serene and solid-looking, and his black frock coat and light gray trousers fitted him snugly and stylishly. They passed along Main Street on the way to church, and the colonel happened to think of a letter on his desk that he wanted, so he told his family to wait at the door a moment while he stopped in his office to get it. He went in and got the letter, and, to his surprise, there was a disreputable-looking pint whisky flask with about an ounce of whisky left in it standing on his desk. The colonel abominates whisky and never touches a drop of anything strong. He supposed that some one, knowing this, had passed his desk, and set the flask there by way of a mild joke.

He looked about for a place to throw the bottle, but the back door was locked, and he tried unsuccessfully to raise the window that overlooked the alley. The colonel's wife, wondering why he was so long in coming, opened the door and surprised him, so that scarcely thinking what he was doing he thrust the flask under his coat tail into his hip pocket.

"Why don't you come on?" asked his wife. "Didn't you find the letter?"

He couldn't do anything but go with her. He should have produced the bottle right there, and explained the situation, but he
neglected his opportunity. He went on down Main Street with his
family, with the pint flask feeling as big as a keg in his pocket. He
was afraid some of them would notice it bulging under his coat,
so he lagged somewhat in the rear. When he entered his pew at
church and sat down there was a sharp crack, and the odor of
mean whisky began to work its way around the church. The colonel saw several people elevate their noses and look inquiringly
around, and he turned as red as a beet. He heard a female voice
in the pew behind him whisper loudly:

"Old Colonel J——is drunk again. They say he is hardly ever
sober now, and some people say he beats his wife nearly every
day."

The colonel recognized the voice of one of the most notorious
female gossipers in Houston. He turned around and glared at her.
She then whispered a little louder:

"Look at him. He really looks dangerous. And to come to church
that way, too!"

The colonel knew that the bottle had cracked and he was afraid to
move, but a piece of it fell out on the floor. He usually knelt
during prayer, but today he sat bolt upright on the seat. His wife
noticed his unusual behavior and whispered:

"James, you don't know how you pain me. You don't pray any
more. I knew what the result would be when I let you go to hear
Ingersoll lecture. You are an infidel. And—what is that I smell?
Oh, James, you have been drinking, and on Sunday, too!"

The colonel's wife put her handkerchief to her eyes, and he
ground his teeth in rage.

After the services were over, and they had reached home, his
wife took her seat on the back porch and began to cap some
strawberries for dinner. This prevented his going out in the back
yard and throwing the bottle over the fence, as he had intended.
His two little boys hung close around him, as they always did on
Sunday, and he found it impossible to get rid of it. He took them
out for a stroll in the front yard. Finally, he sent them both in the
house on some pretext, and drawing out the bottle hurled it into

the street. The crack in it had been only a slight one, and as it struck a soft heap of trash when it fell, it did not break.

The colonel felt immediately relieved, but just as the little boys ran back he heard a voice in the street say:

"See here, sir, law's against throwing glass in the street. I saw you do it, but take it back, and it'll be all right this time."

The colonel turned and saw a big policeman handing the terrible bottle towards him over the fence. He took it and thrust it back into his pocket with a low but expressive remark. His little boys ran up and shouted:

"Oh, papa, what was that the policeman gave you? Let's see it!"

They clutched at his coat tails, and grabbed for his pockets, and the colonel backed against the fence.

"Go away from here, you little devils," he yelled. "Go in the house or I'll thrash you both."

The colonel went into the house and put on his hat. He resolved to get rid of the bottle if he had to walk a mile to do it.

"Where are you going?" asked his wife in astonishment. "Dinner is almost ready. Why don't you pull off your coat and cool off, James, as you usually do?"

She gazed at him with the deepest suspicion, and that irritated him.

"Confound the dinner," he said, angrily. "I'm hungry—no, I mean I'm sick; I don't want any dinner—I'm going to take a walk."

"Papa, please show us what the policeman gave you," said one of his little boys.

"Policeman!" echoed the colonel's wife. "Oh, James, to think that you would act this way! I know you haven't been drinking, but what is the matter with you? Come in and lie down. Let me pull off your coat."

She tried to pull off the colonel's Prince Albert, as she generally did, but he got furiously angry and danced away from her.

"Take your hands off me, woman," he cried. "I've got a headache, and I'm going for a walk. I'll throw the blamed thing away if I have to go to the North Pole to do it."

The colonel's wife shook her head as he went out the gate.

"He's working too hard," she said. "Maybe a walk will do him good."

The colonel went down several blocks watching for an opportunity to dispose of the flask. There were a good many people on the streets, and there seemed to be always somebody looking at him.

Two or three of the colonel's friends met him, and stared at him curiously. His face was much flushed, his hat was on the back of his head and there was a wild glare in his eyes. Some of them passed without speaking, and the colonel laughed bitterly. He was getting desperate. Whenever he would get to a vacant lot, he would stop and gaze searchingly in every direction to see if the coast was clear, so that he could pull out the flask and drop it. People began to watch him from windows, and two or three little boys began to follow him. The colonel turned around and spoke sharply to them, and they replied:

"Look at the old guy with a jag on lookin' for a place to lie down. W'y don't yer go to de calaboose and snooze it off, mister?"

The colonel finally dodged the boys, and his spirits rose as he saw before him a vacant square covered with weeds, in some places as high as his head.

Here was a place where he could get rid of the bottle. The minister of his church lived on the opposite side of the vacant square, but the weeds were so high that the house was completely hidden.

The colonel looked guiltily around and seeing no one, plunged into a path that led through the weeds. When he reached the center, where they were highest, he stopped and drew the whisky flask from his pocket. He looked at it a moment; smiled grimly, and said aloud:

"Well, you've given me lots of trouble that nobody knows anything about but me."

He was about to drop the flask when he heard a noise, and looking up he saw his minister standing in the path before him, gazing at him with horrified eyes.

"My dear Colonel J——," said the good man. "You distress me beyond measure. I never knew that you drank. I am indeed deeply grieved to see you here in this condition."

The colonel was infuriated beyond control. "Don't give a d—— if you are," he shouted. "I'm drunk as a biled owl, and I don't care who knows it. I'm always drunk. I've drunk 15,000 gallons of whisky in the last two weeks. I'm a bad man about this time every Sunday. Here goes the bottle once more for luck."

He hurled the flask at the minister and it struck him on the ear and broke into twenty pieces. The minister let out a yell and turned and ran back to his house.

The colonel gathered a pile of stones and hid among the tall weeds, resolved to fight the whole town as long as his ammunition held out. His hard luck had made him desperate. An hour later three mounted policemen got into the weeds, and the colonel surrendered. He had cooled off by that time enough to explain matters, and as he was well known to be a perfectly sober and temperate citizen, he was allowed to go home.

But you can't get him to pick up a bottle now, empty or full.

Y

EDGAR LEE MASTERS

(A M E R I C A N , 1 8 6 9 – 1 9 5 0)

Deacon Taylor

> I belonged to the church,
> And to the party of prohibition;
> And the villagers thought I died of eating watermelon.
> In truth I had cirrhosis *of the liver,
> For every noon for thirty years,
> I slipped behind the prescription partition
> In Trainor's drug store
> And poured a generous drink
> From the bottle marked
> *"Spiritus frumenti."*†

---

*A degenerative disease of the liver associated with toxicity from alcohol.
†Grain alcohol.

Y

STEPHEN VINCENT BENÉT

(AMERICAN, 1898 – 1943)

Prohibition

"I wouldn't mind if it were gin!" he said,
"Good gin's like ether, sick, with pungent sweet,
And rum I never liked—not even neat.
Champagne and such stuck pins into my head.
Old port was sunlight where a ruby bled.
The silky-bright liqueurs had twinkling feet
Like gipsy children running down a street;
And beer's as old a brother as good bread.

"Still, I could give them up!" he mused and sighed
Like a poor scrawny gust of city wind,
"But it's the precedent that's bad! You'll find
Things worse Hereafter . . . I'd a friend who died.
And . . . well, damned souls had never much to tell . . .
But now they've stopped the Lethe, down in Hell."

CHAPTER SIX

CIDER TO SODA

G. K. CHESTERTON

(ENGLISH, 1874–1936)

A Cider Song
To J.S.M.

EXTRACT FROM A ROMANCE WHICH IS NOT YET WRITTEN AND
PROBABLY NEVER WILL BE

The wine they drink in Paradise
They make in Haute Lorraine;*
God brought it burning from the sod
To be a sign and signal rod
That they that drink the blood of God
Shall never thirst again.

The wine they praise in Paradise
They make in Ponterey,
The purple wine of Paradise,
But we have better at the price;
It's wine they praise on Paradise,
It's cider that they pray.

The wine they want in Paradise
They find in Plodder's End,
The apple wine of Hereford,
Of Hafod Hill and Hereford,
Where woods went down to Hereford,
And there I had a friend.

The soft feet of the blessed go
In the soft western vales,

*The poem compares the wines of France with the fermented apple ciders of West England
and Wales.

The road the silent saints accord,
The road from heaven to Hereford,
Where the apple wood of Hereford
Goes all the way to Wales.

Y

CHRISTOPHER MORLEY

(A M E R I C A N , 1 8 9 0 – 1 9 5 7)

Thoughts on Cider

Our friend Dove Dulcet, the poet, came into our kennel and found
us arm in arm with a deep demijohn of Chester County* cider.
We poured him out a beaker of the cloudy amber juice. It was just
in prime condition, sharpened with a blithe tingle, beaded with a
pleasing bubble of froth. Dove looked upon it with a kindled eye.
His arm raised the tumbler in a manner that showed this gesture
to be one that he had compassed before. The orchard nectar began
to sluice down his throat.

Dove is one who has faced many and grievous woes. His Celtic
soul peers from behind cloudy curtains of alarm. Old unhappy far-
off things and battles long ago fume in the smoke of his pipe. His
girded spirit sees agrarian unrest in the daffodil and industrial riot
in a tin of preserved prunes. He sees the world moving on the
brink of horror and despair. Sweet dalliance with a baked bloater
on a restaurant platter moves him to grief over the hard lot of the
Newfoundland fishing fleet. Six cups of tea warm him to anguish
over the peonage of Sir Thomas Lipton's† coolies in Ceylon. Souls
in perplexity cluster round him like Canadian dimes in a cash reg-
ister in Plattsburgh, New York. He is a human sympathy trust.
When we are on our deathbed we shall send for him. The perfec-

* Pennsylvania.
† British merchant who built the tea empire.

tion of his gentle sorrow will send us roaring out into the dark, and will set a valuable example to the members of our family.

But it is the rack of clouds that makes the sunset lovely. The bosomy vapours of Dove's soul are the palette upon which the decumbent sun of his spirit casts its vivid orange and scarlet colours. His joy is the more perfect to behold because it bursts goldenly through the pangs of his tender heart. His soul is like the infant Moses, cradled among dark and prickly bullrushes; but anon it floats out upon the river and drifts merrily downward on a sparkling spate.

It has nothing to do with Dove, but we will here interject the remark that a pessimist overtaken by liquor is the cheeriest sight in the world. Who is so extravagantly, gloriously, and irresponsibly gay?

Dove's eyes beaconed as the cider went its way. The sweet lingering tang filled the arch of his palate with a soft mellow cheer. His gaze fell upon us as his head tilted gently backward. We wish there had been a painter there—some one like F. Walter Taylor*—to rush onto canvas the gorgeous benignity of his aspect. It would have been a portrait of the rich Flemish school. Dove's eyes were full of a tender emotion, mingled with a charmed and wistful surprise. It was as though the poet was saying he had not realized there was anything so good left on earth. His bearing was devout, religious, mystical. In one moment of revelation (so it appeared to us as we watched) Dove looked upon all the profiles and aspects of life, and found them of noble outline. Not since the grandest of Grand Old Parties went out of power has Dove looked less as though he felt the world were on the verge of an abyss. For several moments revolution and anarchy receded, profiteers were tamed, capital and labour purred together on a mattress of catnip, and the cosmos became a free verse poem. He did not even utter the customary and ungracious remark of those to whom cider potations are given: "That'll be at its best in about a week." We apologize for the cider being a little warmish from standing (discreetly hidden) under our desk. Douce man, he said: "I think cider, like ale, ought not to be drunk too cold. I like it just this way." He stood

* Philadelphia artist and short story writer (1874–1921).

for a moment, filled with theology and metaphysics. "By gracious," he said, "it makes all the other stuff taste like poison." Still he stood for a brief instant, transfixed with complete bliss. It was apparent to us that his mind was busy with apple orchards and autumn sunshine. Perhaps he was wondering whether he could make a poem out of it. Then he turned softly and went back to his job in a life insurance office.

As for ourself, we then poured out another tumbler, lit a corncob pipe, and meditated. Falstaff once said that he had forgotten what the inside of a church looked like.* There will come a time when many of us will perhaps have forgotten what the inside of a saloon looked like, but there will still be the consolation of the cider jug. Like the smell of roasting chestnuts and the comfortable equatorial warmth of an oyster stew, it is a consolation hard to put into words. It calls irresistibly for tobacco; in fact the true cider topper always pulls a long puff at his pipe before each drink, and blows some of the smoke into the glass so that he gulps down some of the blue reek with his draught. Just why this should be, we know not. Also some enthusiasts insist on having small sugared cookies with their cider; others cry loudly for Reading pretzels. Some have ingenious theories about letting the jug stand, either tightly stoppered or else unstoppered, until it becomes "hard." In our experience hard cider is distressingly like drinking vinegar. We prefer it soft, with all its sweetness and the transfusing savour of the fruit animating it. At the peak of its deliciousness it has a small, airy sparkle against the roof of the mouth, a delicate tactile sensation like the feet of dancing flies. This, we presume, is the four and one half to seven percent of sin with which fermented cider is credited by works of reference. There are pedants and bigots who insist that the jug must be stoppered with a corncob. For our own part, the stopper does not stay in the neck long enough after the demijohn reaches us to make it worth while worrying about this matter. Yet a nice attention to detail may prove that the cob has some secret affinity with cider, for a Missouri meerschaum never tastes so well as after three glasses of this rustic elixir.

That ingenious student of social niceties, John Mistletoe, in his

*Henry IV, part 1, III, iii, 9.

famous *Dictionary of Deplorable Facts*—a book which we heartily commend to the curious, for he includes a long and most informing article on cider, tracing its etymology from the old Hebrew word *shaker* meaning "to quaff deeply"—maintains that cider should only be drunk beside an open fire of applewood logs: "And preferably on an evening of storm and wetness, when the swish and sudden pattering of rain against the panes lend an added agreeable snugness to the cheerful scene within, where master and dame sit by the rosy hearth frying sausages in a pan laid on the embers."

This reminds one of the anecdote related by ex-Senator Beveridge* in his *Life of John Marshall.* Justice Story told his wife that the justices of the Supreme Court were of a self-denying habit, never taking wine except in wet weather. "But it does sometimes happen that the Chief Justice will say to me, when the cloth is removed, 'Brother Story, step to the window and see if it does not look like rain.' And if I tell him that the sun is shining brightly, Judge Marshall will sometimes reply, 'All the better, for our jurisdiction extends over so large a territory that the doctrine of chances makes it certain that it must be raining somewhere.' "

Our own theory about cider is that the time to drink it is when it reaches you; and if it hails from Chester County, so much the better.

We remember with gusto a little soliloquy on cider delivered by another friend of ours, as we both stood in a decent ordinary on Fulton Street, going through all the motions of jocularity and cheer. Cider (he said) is our refuge and strength. Cider, he insisted, drawing from his pocket a clipping much tarnished with age, is a drink for men of reason and genteel nurture; a drink for such as desire to drink pleasantly, amiably, healthily, and with perseverance and yet retain the command and superintendence of their faculties. I have here (he continued) a clipping sent me by an eminent architect in the great city of Philadelphia (a city which it is a pleasure for me to contemplate by reason of the beauty and virtue of its women, the infinite vivacity and good temper of its men,

* Albert Jeremiah Beveridge (1862–1927), U. S. senator who sponsored Theodore Roosevelt for president by organizing the Progressive Party. Wrote the *Life of John Marshall*, 4 vols., 1916–1919. Fourth Chief Justice of the United States and founder of the system of constitutional law.

the rectitudinal disposition of its highways)—I have here (he exclaimed) a clipping sent me by an architect of fame, charming parts, and infinite cellarage, explaining the virtues of cider. Cider, this clipping asserts, produces a clearness of the complexion. It brightens the eye, particularly in women, conducing to the composition of generous compliment and all the social suavity that endears the intercourse of the sexes. Longevity, this extract maintains, is the result of application to good cider. The Reverend Martin Johnson, vicar of Dilwyn, in Herefordshire, from 1651 to 1698 (he read from his clipping), wrote:

> *This parish, wherein sider is plentifull, hath many people that do enjoy this blessing of long life; neither are the aged bedridden or decrepit as elsewhere; next to God, wee ascribe it to our flourishing orchards, first that the bloomed trees in spring do not only sweeten but purify the ambient air; next, that they yield us plenty of rich and winy liquors, which do conduce very much to the constant health of our inhabitants. Their ordinary course is to breakfast and sup with toast and sider through the whole Lent; which heightens their appetites and creates in them durable strength to labour.*

There was a pause, and our friend (he is a man of girth and with a brow bearing all the candor of a life of intense thought) leaned against the mahogany counter.

That is very fine, we said, draining our chalice, and feeling brightness of eye, length of years, and durable strength to labour added to our person. In the meantime (we said) why do you not drink the rich and winy liquor which your vessel contains?

He folded up his clipping and put it away with a sigh.

I always have to read that first, he said, to make the damned stuff palatable. It will be ten years, he said, before the friend who sent me that clipping will have to drink any cider.

Ψ

THOMAS HARDY

(E N G L I S H , 1 8 4 0 – 1 9 2 8)

From *The Mayor of Casterbridge*

"Ay. 'Tis Fair Day. Though what you hear now is little more than the clatter and scurry of getting away the money o' children and fools, for the real business is done earlier than this. I've been working within sound o't all day, but I didn't go up—not I. 'Twas no business of mine."

The trusser and his family proceeded on their way, and soon entered the Fair-field, which showed standing-places and pens where many hundreds of horses and sheep had been exhibited and sold in the forenoon, but were now in great part taken away. At present, as their informant had observed, but little real business remained on hand, the chief being the sale by auction of a few inferior animals, that could not otherwise be disposed of, and had been absolutely refused by the better class of traders, who came and went early. Yet the crowd was denser now than during the morning hours, the frivolous contingent of visitors, including journeymen out for a holiday, a stray soldier or two home on furlough, village shopkeepers, and the like, having latterly flocked in; persons whose activities found a congenial field among the peep-shows, toy-stands, waxworks, inspired monsters, disinterested medical men who travelled for the public good, thimble-riggers, nick-nack vendors, and readers of Fate.

Neither of our pedestrians had much heart for these things, and they looked around for a refreshment tent among the many which dotted the down. Two, which stood nearest to them in the ochreous haze of expiring sunlight, seemed almost equally inviting. One was formed of new, milk-hued canvas, and bore red flags on its summit; it announced "Good Home-brewed Beer, Ale, and Cyder."

The other was less new; a little iron stove-pipe came out of it at the back, and in front appeared the placard, "Good Furmity Sold Hear." The man mentally weighed the two inscriptions, and inclined to the former tent.

"No—No—the other one," said the woman. "I always like furmity; and so does Elizabeth-Jane; and so will you. It is nourishing after a long hard day."

"I've never tasted it," said the man. However, he gave way to her representations, and they entered the furmity booth forthwith.

A rather numerous company appeared within, seated at the long narrow tables that ran down the tent on each side. At the upper end stood a stove, containing a charcoal fire, over which hung a large three-legged crock, sufficiently polished round the rim to show that it was made of bell-metal. A haggish creature of about fifty presided, in a white apron, which, as it threw an air of respectability over her as far as it extended, was made so wide as to reach nearly round her waist. She slowly stirred the contents of the pot. The dull scrape of her large spoon was audible throughout the tent as she thus kept from burning the mixture of corn in the grain, milk, raisins, currants, and what not, that composed the antiquated slop in which she dealt. Vessels holding the separate ingredients stood on a white-clothed table of boards and trestles close by.

The young man and woman ordered a basin each of the mixture, steaming hot, and sat down to consume it at leisure. This was very well so far, for furmity, as the woman had said, was nourishing, and as proper a food as could be obtained within the four seas; though, to those not accustomed to it, the grains of wheat, swollen as large as lemon-pips, which floated on its surface, might have a deterrent effect at first.

But there was more in that tent than met the cursory glance; and the man, with the instinct of a perverse character, scented it quickly. After a mincing attack on his bowl, he watched the hag's proceedings from the corner of his eye, and saw the game she played. He winked to her, and passed up his basin in reply to her nod; when she took a bottle from under the table, slily measured out a quantity of its contents, and tipped the same into the man's furmity. The liquor poured in was rum. The man as slily sent back money in payment.

He found the concoction, thus strongly laced, much more to his satisfaction than it had been in its natural state. His wife had observed the proceeding with much uneasiness; but he persuaded her to have hers laced also, and she agreed to a milder allowance after some misgiving.

The man finished his basin, and called for another, the rum being signalled for in yet stronger proportion. The effect of it was soon apparent in his manner, and his wife but too sadly perceived that in strenuously steering off the rocks of the licensed liquor-tent she had only got into maelstrom depths here amongst the smugglers.

Y

EDMUND MILLER

(AMERICAN, 1943 –)

Nutmeg in eggnog
Birthdays the room with New Year's.
Outside . . . icicles.

Y

WILLIAM WANTLING

(AMERICAN, 1933 – 1974)

Lemonade 2¢

Kathy was my
first customer
naturally, I
turned her on
free
she put her

cool hand in
mine
led me to her
dark & sweaty
cellar
kissed me
Lord, how our
lips trembled
how bitter-sweet
& cool
that lemonade

NGUYỄN BINH KHIÊM

(VIETNAMESE, 1491–1585)

A Coconut *

The tree feeds on good soil and drinks up rain.
It flowers and fruits when summer melts down gold.
Inside, the nut wraps round a bowl of ice.
From outside, not a speck of dust slips in.
Better than melon, its fresh milk cools heat.
Like sugarcane, it sobers up a sot.
This water, sent by Heaven, should be shared:
pour out a scoop and slake the people's thirst.

—*Translated by Huynh Sanh Thông*

*According to the translator, the coconut is a symbol of the Confucian scholar who has the capacity to provide good government and sustenance to a thirsty people.

♀

CORNELIA OTIS SKINNER

(A M E R I C A N , 1 9 0 1 – 1 9 7 9)

From *Your Very Good Health*

This lady looked as though, in running a snack bar . . . even a therapeutic snack bar, she had missed her calling. Clad always in peasant blouse, batik scarves and ropes of amber beads, the fortune teller's globe seemed more up her alley than the electric vegetable mixer . . . albeit she lacked the robustness of the conventional Madame Flora. Wan and frail, she looked as if you could knock her over with a camomile leaf. I'd usually find her sitting back of the counter intent on the perusal of a book whose title I never saw but whose content I felt sure was of an occult nature. Seeing me, she would, with great reluctance, put down the volume, having first inserted an Italian leather bookmark. Then, with a soft sigh, she'd rise and with a vague Lady of Shallot* smile would drift in my direction and liltingly enquire, less of me than of some being of the upper air, "May I help you?" My impulse was always to reply "May *I* help *you?*" but instead I'd voice my fancy for a double vegetable juice.

"The special?" she'd sing.

"The special" I'd counterpoint, not having the remotest idea what it meant. With an expression of sweet fortitude, she'd don a pair of rubber gloves many sizes too large for her. Dangling off her bird-like hands, they used to remind me somewhat gruesomely of a clown's false feet. Why she went to such lengths of sanitary precaution, I can't imagine, for she never so much as touched gloved finger to frond of parsley. Taking up a pair of tongs, she would

*Frail, secluded medieval damsel in a poem by Alfred Lord Tennyson; here a pun on a scallion-like vegetable.

gingerly pick the carrots, spinach *et al* out from a square glass tank originally intended for the care and feeding of tropical fish, and deftly place them in the electric dejuicer. Hers was a brisk little machine which, as it ground out the ingredients, gave forth shrill sounds as if the carrots inside it were perishing in excruciating agony. Once, in a crazy moment of thinking she might be of a literary bent and listening to the screams of the expiring vegetables, I jestingly quoted the lines from the Bard about the "shrieks like mandrakes torn out of the ground that living mortals hearing them run mad." . . . which I admit was pretty affected of me. The lady's blank stare indicated she thought I must have heard plenty of mandrakes. I explained it was from *Romeo and Juliet,* whereat she smiled politely and shook her head in a "no-speak-Shakespeare" manner and continued torturing the garden greens.

Y

NORBERT KRAPF

(A M E R I C A N ,   1 9 4 3 –)

Chamomile

Along the border
of an Indiana garden
beside a cold frame

my great-grandparents
cultivated you for
the herb-blossom tea
they believed cured
most of their ills.

Oh calmer of nerves
and delirium tremens,
soother of headaches
and preventer of nightmares,

repeller of insects
and softener of hair

Oh spirit whose steamed
essence unclogged
my infected sinuses
in the Black Forest
and eased my eyelids
toward sleep

may your feathery
foliage and sunburst
flowers flourish in
the herb garden outside
the kitchen window.

♟

FROM *PUNCH*

(C I R C A   1 8 6 0)

Barley Water

AIR—"On the Banks of Allan Water."

For a jug of Barley Water
  Take a saucepan not too small;
Give it to your wife or daughter,
  If within your call.
If her duty you have taught her,
  Very willing each will be
To prepare some Barley Water
  Cheerfully for thee.

For a jug of Barley Water,
  Half a gallon, less or more,

From the filter that you bought her,
  Ask your wife to pour.
When a saucepan you have brought her
  Polish'd bright as bright can be,
In it empty all the water,
  Either you or she.

For your jug of Barley Water
  ('Tis a drink by no means bad),
Some two ounces and a quarter
  Of pearl barley add.
When 'tis boiling, let your daughter
  Skim from blacks to keep it free;
Added to your Barley Water
  Lemon rind should be.

For your jug of Barley Water
  (I have made it very oft),
It must boil, so tell your daughter,
  Till the barley's soft.
Juice of a small lemon's quarter
  Add; then sweeten all like tea;
Strain through sieve your Barley Water—
  'T will delicious be.

Y

JOHN DIGBY

(1 9 3 8 –)

It's a Man's Drink,
Tonic Water

Robert Scarola?—Sure I hated his guts. OK, so I am jealous. He
was taller than me, he was stronger than me, he was smarter than
me, he was more popular than me, he was a better athlete than

me, he was a better student than me, he was better looking than me—Jesus, Joseph, and Mary I was in a no-win situation.

Sure he had charm. He could put the make on the chicks, talk a hungry lion into dropping a lamb out of its jaws. I couldn't even frighten a sparrow away from crumbs scattered on the sidewalk.

Yes I admit it, I was angry, bitter, and hurt. Rivers of green jealousy flowed through my veins. He was the stud, I was the nerd. It seemed that we came from two different planets, yet for some unknown reason we were always thrown together, and believe me I hated it.

I could have forgiven him almost anything—even if he had stolen my mother's milk from out of my mouth. But when he took Marylou away from me, I was smoldering with hatred.

"What went wrong Marylou I waited in the coffee shop for a couple of hours."

"Sorry Rod, I just couldn't make it."

"That's a hell of an excuse."

"Please Rod. I don't want to hurt you."

"You might as well put a dog out of its misery."

"I went out with Robert Scarola instead."

I didn't know what to say. I guess that I could read her mind and it was saying that she found him more interesting than me. He got the chicks, I got the dogs.

It was the same old story; I was always the loser. I remember after the football game the way he stood in the locker room with his strong legs apart, his towel wrapped around his slim waist, combing his wet hair and saying (knowing that I knew he had stolen Marylou away from me), "I've got no trouble with girls."

Those green rivers of jealousy were bubbling up in my veins. I wanted to belt the sucker. But what was the use? He would have belted me back and once again I would have come off second best.

He was the better man. Christ, even if I studied like crazy around the clock, I would get a B-plus while he hardly opened a book and always got an A.

There was no doubt, he was a first-class citizen, I felt second class.

We had nothing in common but one thing: We both loved and guzzled quarts of tonic water. Sure it was strange, but each to his own. We never drank anything else. Other beverages held no interest for us. As soon as my mother tore me from her breast I was drinking tonic water. Honest to God, I drank nothing else. How I came to the stuff I can't recall. But even then I couldn't win. The day I ordered a tonic water at the pizza place Scarola happened to be sitting at the booth in front of me.

"Trying to emulate your betters, Sullivan?" he said.

A couple of chicks from his permanent harem giggled. I blushed and walked away without touching my drink. Believe me I felt low, a worm stood higher than me that day. For the next few weeks I let my grades drop. I was hell-bent on revenge. But how? I just wanted a little satisfaction.

My opportunity came in a strange way.

Our high school was having its senior survival course: four teams, blue, red, green, and brown. Each group consisted of about twenty guys. The four groups were bundled into a covered truck so that we had no idea where we were taken, and dumped down somewhere in the middle of nowhere, probably about fifty or sixty miles away from school. The first group to reach school graduated with a gold medal. Each guy was allowed a backpack, a pocket knife, food, and water. The groups were dumped at various points. I was assigned to blue group and of course Scarola was the group leader.

Scarola and I took our own bottles of tonic water.

We were the last group to be dumped in the middle of nowhere. Scarola jumped out of the truck, looked at the sinking sun in the late afternoon, and pointed north.

"This way guys. Follow me and we'll be the first back."

I don't think I was the only one who hated his guts.

The object of the operation was to reach school by the next early evening. If any of us failed, the school bus would come search for us, and pick up the stragglers. Scarola had no thought of failing. His group was going to win.

How many miles did we walk? How many rocks did we climb? How many rivers did we wade across? All evening he pushed us forward until it was either too dark to see a hand in front of us or we just dropped from exhaustion.

"OK guys. We can rest for the night," he offered, as he threw himself down on a bed of soft pine needles at the edge of the woods.

"Tomorrow by my reckoning we ought to be back way before the others."

Everybody ignored him and were intent on resting or eating before he gave another order.

Was it by accident or design that I dropped down by the side of him with our backpacks between us?

"You're a lucky guy to be in my group, Sullivan," he said, rolling over to sleep. "With me, you'll be among the first back."

I think he said it to anger me.

I stretched out on my back and was about to fall into a deep sleep when the idea suddenly hit me. Sure it was disgusting, but revenge is sweet.

Scarola was soon out cold. I made sure that no one could see me in the dark. Then I reached into his backpack, lifted out his bottle of tonic water, and crept away into the woods.

OK. So it was disgusting. I unscrewed the bottle top, poured half away . . . and Jesus, Mary, and Joseph I did it! I peed into his bottle. I almost broke my fingers, thumb, and wrist screwing the top back on again.

The following morning he got us up when the stars were still hanging in the sky. I'll tell you I was pretty excited. I just wanted to see his face when he took a gulp of the tonic water.

"Save your water guys, we've got a long push this morning." We all obediently obeyed him. Jesus was I disappointed.

For miles we must have tramped. That bastard wouldn't rest, wouldn't stop to drink. Finally, a little, after lunch, we all threw ourselves on a patch of wet, uncut grass and refused to go on any farther without first having something to drink.

"Guys," he said, addressing us as if we were his own personal minions. "Let's eat and drink."

O Lord, revenge is sweet. I sat in front of him anxiously waiting for the blessed moment.

"It's a man's drink, Sullivan," he said as he unscrewed his bottle of tonic water.

I was waiting. God, did I try not to show any emotion on my face. He drank a long draught. And nothing happened. To say that I was crestfallen was an understatement. He took another big gulp.

"Great stuff, Sullivan, as good as mother's milk."

I couldn't believe it. I had failed again. He had the better of me again. I unscrewed my bottle of tonic water and with the first mouthful realized my mistake. In my excitement and the darkness I must have peed in my own bottle. It was a mistake anyone could have made.

As I swallowed the tonic water my face must have distorted into a mask of absolute pain and horror.

"What's wrong, Sullivan, can't you take it?"

My insides twisted into a knot of rage and the bitter taste of my own urine lingered on my tongue.

"Sure I can take it. I love the nectar."

I was determined not to show that anything was wrong. I drank half a bottle of the damn stuff. God, it was awful, it was hell. A little later I crawled away while he was not looking and vomited my guts out.

I never knew if he had discovered my game and switched bottles while I was asleep or I had made a genuine mistake that night.

We were the first group to arrive back at school.

I never touched tonic water again for over twenty years.

After that I tended to avoid him out of shame and defeat. We both graduated. He went to Yale and I believe he studied Liberal Arts. I went to a small east-coast college and took a degree in business management.

I lost touch with him. To tell you the truth, I forgot all about him until more than twenty-five years later when my personal secretary brought me a letter from a small town in the midwest.

It was a fawning letter written by the mayor. I was planning to open another fast-food restaurant just outside of the town where a new highway was coming through. The letter rambled on telling me how much Elton would appreciate having Frytown in their vicinity as it would open up employment after a long slump. The letter was signed Robert Scarola.

I asked my secretary to check the guy out. Obviously he never knew that I was the owner of a chain of restaurants that stretched from coast to coast.

A little later that week my secretary brought me the relevant information. It was the same guy, there could only be one Robert Scarola.

"The most likely man to succeed" didn't graduate from Yale; he left without taking a degree, drifted into real estate, and ended up as mayor of a one-horse town going no place.

I called my secretary and asked her to arrange a meeting with Mr. Scarola. I guess it was out of curiosity that I wanted to touch base again with the golden boy.

"Can't be done, Mr. Sullivan," my secretary said.

"Can't be done?"

"Mr. Scarola died over a month ago, a heart attack I believe. Mr. Forester, his deputy, is mayor now."

"He was the same age as me," I offered.

"Sorry, Mr. Sullivan, I don't know about that. Will that be all?"

"Yes, of course," I muttered vaguely.

"You have a two o'clock meeting with Mr. Jeong, and your wife called to remind you that the guests are expected at six-thirty for cocktails."

I wasn't even listening. She stood there for a few seconds and then she must have thought better of it and departed, closing the office door behind her.

For the next few minutes I couldn't even begin to sort out the memories that came rushing through my head like an express train. Everything came up like waves from the past.

That night I couldn't sleep and around three o'clock I got up from bed, left my wife still peacefully sleeping after our dinner party, and went to the fridge. I was dry and wanted a drink. I opened the door and on the top shelf was a bottle of tonic water. I reached in and held the bottle in my hand. It was cold to the touch. I hadn't tasted the stuff for over twenty-five years.

As I unscrewed the top I thought of a similar instant during the survival course back in high school and of Robert Scarola. I raised the bottle in the air. Maybe he wasn't so great, maybe he wasn't the better guy than me . . . maybe he backed the wrong horse. I said in rather a too-loud voice, "Here's to you, Robert Scarola, let's hope you're among the gods drinking tonic water with plenty of chicks."

There was no bitterness, no envy, no jealousy, no remorse, no regrets. I took a long drink, my first in a long time. It tasted good, believe me, it really tasted good.

♼

FRANK MOORHOUSE

(A U S T R A L I A N , 1 9 3 8 -)

The Coca-Cola Kid

"I don't know why we're here," she said.

They slammed the Citroën doors in unison.

"You know why we're here," he said.

They walked across the lawn to the door of the Yacht Club. She didn't answer so he answered for her ". . . to meet the enemy," he said, as they went from the chirping country darkness into the rattle.

The Young Liberals were not yet drunk. A few couples had begun to dance but were not yet loose. All the poker machines were in use. The band was playing—staring out, in a trance, to the private world of dance musicians. The stewards swayed with trays of drinks, rushing with a professional slowness. The dance music and intermittent slugging of the poker machines filled the crevices between conversations.

They left their coats at the door.

"Perhaps I should've worn a dinner suit," he said, not meaning it. Trying to fill the dull emptiness between him and his wife.

"Only a few're in dinner suits," she said.

He knew why they were there. They were there because they were bored and lonely in a town where they hardly knew anyone—let alone enough people to pick and choose politically. The town didn't even have a Communist Party.

He looked for a drink, to ease himself into the ill-fitting situation. A shoehorn drink.

He stopped a steward with one hand and turning to his wife said, "What'll you have?"

"Beer," she said.

He wanted to throw a few down fast.

"There's Kevin and Gwen," she said, pointing her beer across the dance floor.

"The revolutionaries made it," Kevin said, greeting them.

"We're waiting to be thrown out," he said.

"Please—no politics tonight," Kevin said laughing. "Remember I arranged the invitation."

Kevin introduced them to another couple. They talked. Then he danced with Sylvia. "Now you've brought us here, at least be agreeable," she said to him. He felt as soon as she said it that the party was a mistake.

After the dance he drifted off and lost four dollars on the poker machines and resented it.

"I warned you to limit yourself to two dollars," his wife said when he drifted back to where Kevin, Gwen and she were seated around a jug of beer.

He wasn't looking for chastisement. He glanced at her but didn't retort.

He began an aggressive talk with Percival who was on the executive of the group—one of the political ones.

"I agree," said Percival. "Take the farms away from those who mismanage them."

His aggression was seeped away by the agreement.

"I'm pleased you agree," he said, not meaning it, with it sounding so agreeable that it disgusted him. He was caught in polite conversation. He wanted to be disagreeable. He wanted to demonstrate where he stood, who he was. He wasn't one of them.

He caught his wife's expression which said, "Must you talk politics?" and looked away.

He stood up and left the table. He was alone again. He resisted the temptation to go back to the poker machines. But he didn't want the small talk of the table. He didn't want the bitchiness of his wife. He stood alone, a hand in his pocket, drinking too fast.

He went to the balcony of the club. It was then that he met the American.

He was standing alone leaning against a wall, drink in his hand, bow-tied, looking across the water. He turned and acknowledged him with his eyes and raised his glass in salute, saying in a slightly slurred American accent,

*"I must go down to the seas again to the lonely sea and the sky. All
I ask is a tall ship and a star to steer her by."*

A smile squeezed out of him involuntarily as he heard the
American's recitation—it disarmed him.

"And furthermore, buddy, I know the last stanza too," the
American said.

"Must go down to the seas again to the vagrant gypsy . . . da da
dah da dah . . . *and all I ask is a merry yarn from a laughing fellow
rover."* The American seemed to falter, paused, and then continued, *"and a quiet sleep and a sweet dream when the long trick's over."*

The American's face then sagged, sadly. He drank from his glass,
his hand wrapped around it in a tight grip.

He said to the American, "I thought only Australian and British
kids learned Masefield."

"I learned Masefield, buddy, I learned Masefield." His face
tightened up again for performance,

*"I'm going to be a pirate with a bright brass pivot gun,
And an island in the Spanish Main beyond the setting sun."*

"I don't know that one."

"The Tarry Buccaneer," he said, "The Tarry Buccaneer."
He lapsed again.

"Do you sail?" he asked the American, again writhing against
the compulsion to be "sociable."

"I fly," the American said, and then in a formal voice, "So you'd
like to be a pirate with a big brass pivot gun—come in, sit down—
do you smoke?—no, have one of mine—I think we can find you a
suitable position—provided of course that you have the requisite
qualifications—have you a brass pivot gun?" It was a meandering
performance for the American's private amusement.

Then the American turned to him as though he had remembered his presence and became genial. "My name's Becker—Atlanta—the Coca-Cola kid from Coca-Cola City."

They shook hands heavily and overlong like two drunks. "How
do you mean Coca-Cola kid?"

"I'm the Coca-Cola kid—Coca-Cola—work for Coca-Cola—here
from the parent company."

His resistance to Becker grouped instantly, his sociability strong-
armed out of the way. An enemy.

"Out here to rot some teeth," he said, softening it only mildly with a quarter smile, glad at last to be critical, although aware of the effort it took to be unpleasant against the mood.

Becker smiled, puzzled, and seemed to fumble to make out the true nature of the remark, as though only hazily aware that it was not just facetious.

He wished he'd been able to say something that let Becker know he'd met the unexpected—that he was face to face with a radical.

"Hey, hey, hey," Becker said, coming to a realisation, almost singing, "what have we here—an enemy of Coca-Cola?"

"I guess I am," he said, trying to be firm—it wasn't exactly how he'd seen himself—"among other things."

Becker shook his head with disbelief. "Could not be—could not be—the enemies of Coca-Cola have all been liquidated—Atlanta had them all liquidated."

He didn't smile at the humour.

"How could you talk about the Holy Water like that?" Becker said. "Why, we made a profit of eighty-nine million dollars last year—why, baby, the whole world drinks Coca-Cola."

"A few million people can't be wrong," he said, coldly.

"A few million? Baby, ninety-five million people drink Coke every single day—you saying people don't know what they like?"

Becker saw a steward. "Steward, my good man—" he turned to him and asked what he was drinking and then ordered.

"Why, Coca-Cola," Becker went on, "is an everyday word in every language—it's in the dictionaries."

"Truck drivers wash their engine parts in it to eat away the grease," he said, "and stomach linings and teeth. . ." He wanted to say something about economic priorities.

"Their parts?—I wash my parts in it too. Don't tell me the one about the tooth dissolving overnight in a glass of Coke—I don't want to hear that one." He shook his head. The steward came back with the drinks.

"What do you do for the good of the world, Kim boy?" Becker drank deeply.

"I'm a teacher," he said, feeling a certain virtue.

"Yes," Becker said, looking out to sea, "you teach good habits, I teach the bad."

"If that's the way you feel," he said, seizing the point, "why don't you get out?"

Becker turned back with a grin. "Hell, I'm not serious—I like Coke—hell!"

He blushed at having failed to see Becker's flippancy.

"Now look—" Becker said, with a loud selling voice, "you might give kids a lot of good ideas but you give them a lot of God-almighty crap too—I just push a good soft drink—the best as it so happens—nothing more, nothing less."

"But the waste of it all." He moved on to the line he wanted to argue. "The waste of resources when they could be used for food. Socially desirable ends."

"I noticed you spending a few damn dollars on social undesir-ables," Becker said, clinking their glasses to make his point.

"Personal charity is no answer."

"Stop." Becker held up his hand in a halt sign. "Stop—the game's over. I'm here to get drunk, not to sell Coca-Cola or debate the world."

He wondered if he'd scored against Becker. It was typical of people like Becker that they could switch off matters of con-science.

Becker had gone for more drinks.

He came back. "I'm going to be a pirate with a bright brass pivot gun." Gracefully Becker presented the drink. "I want to tell you a secret." Becker sounded drunk. "I don't want to be the Coca-Cola kid," he shook his head, "I want to be a jazz pirate."

"Jazz pirate?"

"Jazz pianist—jazz penis," Becker said.

"Why don't you?" he asked without interest. Becker was ob-viously not going to talk seriously.

"They don't have jazz penises any more—not any more." He waved his drink around. "I practised every day—I wrote for les-sons—filled in a coupon back of a comic book."

At first he thought Becker was rambling on with nonsense but realised he wasn't altogether joking. He looked at him with slightly more sympathy, although jazz pianists weren't actually in the van-guard of social reform. Decadent white jazz.

He stood there to finish the drink with Becker, who didn't say any more. He seemed to be conversing with his own thoughts.

He found himself with nothing to say either and stood for a second rebelliously uncomfortable—stymied—unable to engage Becker. Then he wandered off, with a glance at Becker, who wasn't looking and appeared not to notice.

He saw his wife dancing with Kevin. They weren't aware he was watching. He felt as if he were spying. He was spying—but his wife was innocently serious and there was a quite proper distance between them. They were talking more than dancing. He joined them when the dance finished.

"Where've you been?" his wife queried.

"Talking to an American—the Coca-Cola Kid."

"What's he doing here?"

"Reciting Masefield."

Her face expressed irritation and she turned to Kevin, obviously with no intention of forcing her way through his enigmatic answers. He didn't care to help.

Most people were on the way to being drunk. The poker machines were rattling faster. The dancing was full of sex.

"Enjoying it?" he heard his wife say politely to him as though she felt it necessary to say something to link herself with him, perhaps as a conciliatory offer. He recognised it as a move to end the bad day they'd been having together.

"What can you expect from the sons and daughters of graziers and Coca-Cola executives?" he said, matching the effort of her wifely question with a husbandly tone. But it really didn't dissolve the bad feeling between them. Somehow he didn't care, somehow he preferred the bad feeling.

Later he went to the lavatory. In the lavatory standing side by side at the urinal he heard two young men talking about skiing. "She had the bottom part of her leg in plaster, you see," one said, and the other laughed.

He felt distaste and resentment. Because of their exclusiveness.

He decided to look at the other rooms of the club. Along a corridor he heard piano playing. Sloppy piano playing. He followed the sound into a small carpeted room with glass cases containing a few books fallen like dominoes. A few dusty trophies. Over in the far corner in the half light Becker was playing the piano, his drink on the piano seat beside him.

He felt obliged to say something now he'd entered the room.

"Practising for that job as jazz pianist?" he said.

Becker looked at him but didn't say anything. He kept on playing, slopping his head with the tune. His bow tie seemed to fit the part.

He stood and listened. He realised that Becker wasn't going to talk while he was playing. He drifted over to the books—yachting records. Joshua Slocum. Peter Heaton.

Becker stopped playing and drank from his glass.

"What do you know about jazz, Mr. Teacher?"

"Very little—the Negro struggle. . ."

Becker didn't say anything in reply.

He felt dismissed. He felt he'd been assessed and dismissed.

Becker began to play again.

He stood there half listening to the playing, but Becker's eyes didn't return to him, didn't further acknowledge his presence. Becker sang in a deep imitation of a negroid voice, snatches of song, but not for him, not for any audience. Occasionally Becker would look in his direction but not at him.

He became aware of his own body, his feet, legs, torso, hands, arms, neck and head, standing there in an upright position on a floor, in a club, in a town, on an orbiting planet—standing in stark isolation as the world orbited the sun. He didn't belong with Becker's playing. He didn't belong with the trophies. Or with the crowd outside. Even his wife—he was not in contact with his wife.

He listened to the playing but always thoughts about himself pushed in between him and the music. Not that the music was easy to listen to. It broke and fumbled with Becker doggedly retracing his fumbling.

He stood there, not having the urge to do anything else but stand. Realising himself unacknowledged by anyone.

His thoughts were not anything more than pings of discomfort and a rasping uneasiness. He desperately wanted to be pleased with himself. Standing at the Yacht Club with Young Liberals dancing around him, out of place, argumentatively drunk and no one to argue with. He felt distinctly displeased with himself.

He wanted to assert himself with Becker. With Becker especially. Because he was American. For some reason he felt free to say anything to Becker, no holds barred. But Becker had eluded him. He could not snare Becker.

Becker's and his eyes met. Becker held the stare and sang at him and then looked back to the keys. Becker had sung the song against him.

He felt he had to go. Again he didn't know whether to try to indicate goodbye to Becker. He watched for an opportunity but it didn't come and the playing stopped him from saying anything.

He gave a slight shrug and left the room.

Just outside the door he heard Becker say, "I didn't want to be a jazz pianist. Or a brass pivot gun. Because I'm the Coca-Cola Kid."

He stopped. So Becker had been aware of him. But nothing came to his mind to say to Becker. Despite an aggressiveness the statement was full of some sort of appeal. It deflected attack. But he couldn't respond because its special appeal was too subtle, and he was too blurred with drink. He was opposed to Becker, but he wasn't game enough to risk being wrong or embarrassed with Becker. So he walked on. Becker would think he hadn't heard.

Out in the bright noise he was able to merge with the crowd and did not feel as intensely isolated.

He played another dollar in the poker machines and won sixty cents.

He wasn't socially acceptable or socially adroit and he didn't claim to be or try to be. So what? There was something phony about a "good mixer"—a good mixer had to smother his real reactions. He didn't have time for niceties or phoniness. Other things were more important.

After a while he moved back to his wife and the others. They chatted.

His wife said, "You've been great company tonight." The wound opened again.

Then the band played the anthem. He hadn't realised it was so late.

"Have a drink for the road," Kevin said, "if you're in our party." Friendly sarcasm.

"Where have you been all this time?" his wife said, annoyed.

"Looking over the club—listening to the American playing the piano."

"The mythical American—was he reciting Kipling again?"

"Masefield."

"Why didn't you bring him over if he was so captivating?"

"He was drunk."

"You're *not?*"

The crowd was a criss-cross of unravelling knots. The band was shaking spittle from their instruments, unscrewing mouthpieces.

A sudden movement attracted his eyes. A young man in a dinner suit with thick blond hair had changed from a party-goer into an official. He was walking urgently through the crowd with a steward. They went through the doorway to the men's lavatory. Outside in the hall, on the edge of the unravelling crowd, a couple of stewards had taken off their bow ties and they were stacking chairs.

Kevin came over with the drinks and as he handed them around he said, "A fellow tried to hang himself in the men's—they cut him down in time."

He sensed it was the American and tightened.

His wife said, "He's still alive?"

"Yes—just—an American—" Kevin turned to him. "Must be your American."

"The Coca-Cola Kid—Becker." He was galvanised. For the first time that night he felt lifted out of his isolated preoccupation.

"The one you were talking to?" his wife said, her voice loud with the shock of it.

"He was playing the piano about half an hour ago—I was listening to him."

The crowd had smelt the event and were looking towards the lavatory. Murmuring questions.

"He was blue in the face," Kevin said. "They're rushing him to hospital. A steward told me he was quite a mess when they cut him down."

They drank their last drinks watching the lavatory but the American was not carried out.

"Must have taken him out the back way," Kevin said.

"He told me Coca-Cola was a common noun," he told them, and being English teachers they smiled.

They moved towards the door. "He didn't seem suicidal—just drunk," he told them.

Outside they separated from Kevin and Gwen. He said to his wife, "Seemed just another genial American," and then added, "the man who presses the bomb release, lynches the Negro, drops the napalm is just another genial American—good fun at cocktail parties."

His wife didn't comment.

As he started the car he said, "I suppose working for Coca-Cola would be enough to make you want to kill yourself."

"Come off it," his wife said, irritably.

"A victim of a personality destroying system," he said, niggling her.

"Must you always be so doctrinaire?"

"I hope he dies," he said, goaded further, driving hard over the cattle grid at the gate.

"Kim!" she turned towards him, "don't be so heartless."

"You're getting soft," he said. "How many Asians starve to death so the Americans can drink Coca-Cola?"

He swung hard but hit a bad bump.

An empty bottle on the floor of the car rolled wildly under their feet.

"Do you want me to drive?" she said, slightingly.

He ignored it.

"Things aren't as simple as you sometimes see them," she said. Then, lighting a cigarette, she laughed derisively. "You'd have fainted if they'd carried the body out."

"Lenin never watched executions," he said, accelerating on the straight road towards the scattered light of the town.

He saw her snort and look out her window across the corn fields, turning completely away from him.

ROBERT HOLLANDER

(AMERICAN, 1933 –)

You Too? Me Too—Why Not? Soda Pop

I am
look
ing at
the Co
ca Cola
bottle
which is
green wi
th ridges
just like
c   c   c
o   o   o
l   l   l
u   u   u
m   m   m
n   n   n
s   s   s
and on itself it says

COCA-COLA
reg.u.s.pat.off.

exactly like an art pop
statue of that kind of
bottle but not so green
that the juice inside
gives other than the co
lor it has when I pour
it out in a clear glass
glass on this table top
(it's making me thirsty
all this winking and
beading of Hippocrene
please let me pause
drinking the fluid in)
ah! it is enticing how
each color is the same
brown in green bottle
brown in uplifted glass
making each utensil on
the table laid a brown
fork in a brown shade
making me long to watch
them harvesting the crop
which makes the deep-aged
rich brown wine of America
that is to say which makes
soda   pop

CHAPTER SEVEN

COFFEE, TEA, OR MILK? A LITTLE COCOA PERHAPS!

HENRY CAREY

(ENGLISH, ?1687–1743)

An Ode in Praise of Coffee

Thou sacred liquour of nectarous taste,
O Coffee! thou whose fame shall last
While o'er thy glowing charms we smile,
And bless the product of thy parent soil;
When spleen and all its direful train
Despotick vex my tortur'd brain,
Thy presence makes the bold usurpers fly,
Leaving my peaceful soul possess'd of calm serenity.

Charm'd with thy scent, and by thy help reviv'd,
When of my reason I'm deprived
By the prevailing charms of wine,
Or glorious Punch, as equally divine,
Thy healing virtues soon destroy
Whatever can my sense annoy.
But who, Ah! who can tell how those are blest
Who share the pleasures thou bestow'st to crown a
plenteous feast.

❦

SAMUEL BUTLER

(E N G L I S H ,   1 6 1 2 – 1 6 8 0)

From *Characters*

A COFFEE-MAN

Keeps a coffee market, where people of all qualities and conditions meet, to trade in foreign drinks and newes, ale, smoak, and controversy. He admits of no distinction of persons, but gentleman, mechanic, lord, and scoundrel mix, and are all of a piece, as if they were resolv'd into their first principles. His house is a kind of *Athenian* school, where all manner of opinions are profest and maintain'd to the last drop of coffee, which should seem, by the sovereign virtue it has to strengthen politic notions, to be, as some authors hold, the black porrige of the *Lacedemonians,** and the very same *Lycurgus* himself us'd when he compos'd his laws, and among other wholesome constitutions hit upon that, which enjoins women to wear slits in their petticoats and boys to steal bread and butter, as *Plutarch* writes in his life. Beside this their manner of conversing with strangers and acquaintances, all in one company, agrees perfectly with the custom of the *Spartans,* that made their city but one family, and eat and drunk all together in public. He sells burnt water and burnt beans as puddle, and of as pure a race, and though not altogether so delicious upon the palate warmer in the stomach, that never stirs the blood with wanton heates, nor raises idle fancies in the brain, but sober and discreet imaginations, such as black choler, like it self, produces. It is a kind of drink, as curses are a kind of prayers, that neither nourishes, nor quenches thirst. *Dives*† would hardly endure a drop of it on the tip of his tongue. He is a *Barbarian* brewer of *Mahometan Taplash,*‡ that tempers his

* Spartans.
† Generic name for a rich man, from Luke 16:15.
‡ Dregs.

decoction according to the *Alcoran, and skinks in earthen goblets to his guests.* If it were not for news and the cheapness of company he would be utterly abandon'd: for that, with the freedom to vapour, lye, and loiter upon free cost, draws more company than his coffee, or the *Turk* that drinks it on his sign,* though that be the better of the two. Coffee, though the vilest of liquors, carries away the name of the house from chocolate and tea drinks of better quality, that are equally sold there and of better reputation, even as mean thieves are only call'd so, and great ones taken no notice of.

Y

HISAHI ITO

(J A P A N E S E , 1 9 2 3 –)

Coffee/No Sugar

Shop door opens:
two men in,
a little smoke out.
"Lay him out
by the fourth round. . . .
I'll have beer.
Coffee for you—right?"

Getting dark; bar
signs begin to flicker
in the evening,
in the gentle hearts
of men, my kind.
Am I looking
at the new champ?
Above him and his coffee

* Shop sign of a coffeehouse.

a poster pronounces
time and place.

"Remember your right, boy.
Your right": the fat one,
gulping beer.
The young one is
silent, taking his coffee
black/no sugar.
His "right" barely moves,
quick/sure, when he drinks.

I pay: go out: leave
the fighter's world.
A fine May evening,
floating, refreshing.
Crossing the tracks
I pace slowly
towards the arena.

—*Translated by Thomas Fitzsimmons*

GIUSEPPI BELLI

(ITALIAN , 1 7 9 1 - 1 8 6 3)

The Coffeehouse Philosopher

People differ the way one coffee bean
in a sack will differ from another
when they're spooned into some expresso machine
to meet the fate they all come to together.

They go round and round, behind, before,
always changing places; they've barely begun

when there they go through that iron door
that crushes them into powder, all one.

So all the people live on this earth,
mixed together by fate, swirled around
and changing places, bumping along from birth,

not knowing or caring why, some out of breath,
some taking it easy, all sinking toward the ground
to be gulped down by the throat of death.

—*Translated by Miller Williams*

Y

BERT MEYERS

(AMERICAN, 1928–1979)

Now It's Friday

I came for coffee
to water my deep heart

Now it's Friday
and my hands still hurt
from Monday Tuesday

But a cup of coffee
is a big brown eye
that looks at anyone

Where is the door
that opens like a hug

When you're always alone
at night there are the stars
The sky's a plate of salt

And like a loaf of bread
You wait growing hard

Y

ROBERT BENCHLEY

(A M E R I C A N , 1 8 8 9 – 1 9 4 5)

From *My Ten Years in a Quandary*

COFFEE VERSUS GIN

What is it with these people at Cornell and other hotbeds of medical research that they are always monkeying around with experiments on liquor? They are always trying to find out how many cubic centimeters of alcohol you can take before the salivary glands start drying up, or how much black coffee it takes to counteract the effects of a shot of gin. What do you *care*?

Presumably drinking is a carefree occupation; at any rate while you are drinking. If it isn't a lark, it certainly isn't anything else. It certainly isn't practical; we know that. I don't suppose you would find one man in a hundred who has made a nickel by taking a drink.

So why all this "cubic centimeter" talk? Why go at it so hard-headedly? Why drag in the salivary glands? *We* know they dry up. You don't have to work in a Cornell laboratory to know that.

The particular experiments which have thrown me into such a fever-heat of indignation had to do with the use of coffee as an antidote for liquor. Seventy-five cubic centimeters of gin followed

immediately by ten grams of coffee in a half pint of water and the gin had no effect. I've got a better scheme than that. Don't take the seventy-five centimeters of gin at all. Think of all the coffee you'd save!

And, incidentally, I consider coffee greatly overrated as a stimulant. It has never kept me awake yet and it has never started me off with a bang in the morning. A lot of people say: "I'm no good in the morning until I've had my coffee." I'm no good in the morning even *after* I've had my coffee.

This old-wives' superstition that a cup of black coffee will "put you on your feet" with a hangover is either propaganda by the coffee people or the work of dilettante drinkers who get giddy on cooking-sherry. A man with a *real* hangover is in no mood to be told "Just take a cup of black coffee" or "The thing for you is a couple of aspirin." A real hangover is nothing to try out family remedies on. The only cure for a real hangover is death.

On such rare occasions as I feel called upon to work late at night, a cup of black coffee taken at midnight acts as an instantaneous soporific. Two cups and I oversleep in the morning. I like coffee, but it soothes me. And that is one thing I don't need—soothing.

The same people who tell you that a cup of black coffee will put you "on your feet" are also the ones who go around recommending a "good dose of castor oil" for a broken leg. (Why must it always be a *good* dose of castor oil? There is no such thing as a "good" dose of castor oil.) They tell you how to cure hiccoughs, and swear by a glass of hot milk in cases of insomnia. They are nice, kindly people, but you will usually find that they lead fairly sheltered lives. They don't get around much in real suffering circles.

And Cornell or no Cornell, I still don't believe that ten grams of coffee in a half pint of water will offset seventy-five cubic centimeters of gin. How much is seventy-five cubic centimeters of gin, anyway?

♀

STANLEY J. SHARPLESS

(ENGLISH, 1910-)

In Praise of Cocoa, Cupid's Nightcap

LINES WRITTEN UPON HEARING THE STARTLING NEWS THAT COCOA IS, IN FACT, A MILD APHRODISIAC

Half past nine—high time for supper;
"Cocoa, love?" "Of course, my dear."
Helen thinks it quite delicious,
John prefers it now to beer.
Knocking back the sepia potion,
Hubby winks, says, "Who's for bed?"
"Shan't be long," says Helen softly,
Cheeks a faintly flushing red.
For they've stumbled on the secret
Of a love that never wanes,
Rapt beneath the tumbled bedclothes,
Cocoa coursing through their veins.

JOHN UPDIKE

(AMERICAN, 1932-)

Lament, for Cocoa

The scum has come.
  My cocoa's cold.
The cup is numb,
  And I grow old.

It seems an age
  Since from the pot
It bubbled, beige
  And burning hot—

Too hot to be
  Too quickly quaffed.
Accordingly,
  I found a draft

And in it placed
  The boiling brew
And took a taste
  Of toast or two.

Alas, time flies
  And minutes chill;
My cocoa lies
  Dull brown and still.

How wearisome!
  In likelihood,
The scum, once come,
  Is come for good.

Y

PHILIP FRENEAU

(AMERICAN, 1752–1832)

The Dish of Tea

Let some in beer place their delight,
O'er bottled porter waste the night,
   Or sip the rosy wine:
A dish of Tea more pleases me,
Yields softer joys, provokes less noise,
   And breeds no base design.

From China's groves, this present brought,
Enlivens every power of thought,
   Riggs many a ship for sea:
Old maids it warms, young widows charms;
And ladies' men, not one in ten
   But courts them for their Tea.

When throbbing pains assail my head,
And dullness o'er my brain is spread,
   (The muse no longer kind)
A single sip dispels the hyp:
To chace the gloom, fresh spirits come,
   The flood-tide of the mind.

When worn with toil, or vext with care,
Let *Susan* but this draught prepare,
   And I forget my pain.
This magic bowl revives the soul;
With gentlest sway, bids care be gay;
   Nor mounts, to cloud the brain—

If learned men the truth would speak
They prize it far beyond their Greek,
  More fond attention pay;
No Hebrew root so well can suit;
More quickly taught, less dearly bought,
  Yet *studied* twice a day.

THÉODORE DE BANVILLE

(FRENCH, 1823–1891)

Tea

Miss Ellen pour me some tea
In a pretty Chinese cup
Where fighting mad a goldfish
Taunts a monster from the sea.

I love the wild cruelty
Of the chimeras one shuts up:
Miss Ellen pour me some tea
In the pretty Chinese cup.

Under a sky of red immensity
A proud and cunning lady,
With her turquoise eyes so shady,
Shows ecstasy and naivete.
Miss Ellen pour me some tea.

—*Translated by John Digby*

Y

LEIGH HUNT

(E N G L I S H , 1 7 8 4 - 1 8 5 9)

Tea-Drinking

The very word *tea,* so petty, so infantine, so winking-eyed, so expressive, somehow or other, of something inexpressibly minute and satisfied with a little (*tee!*), resembles the idea one has (perhaps a very mistaken one) of that extraordinary people, of whom Europeans know little or nothing, except that they sell us this preparation, bow back again our ambassadors, have a language consisting only of a few hundred words, gave us *China*-ware and the strange pictures on our tea-cups, made a certain progress in civilisation long before we did, mysteriously stopped at it and would go no further, and, if numbers and the customs of "venerable ancestors" are to carry the day, are at once the most populous and the most respectable nation on the face of the earth. As a population they certainly are a most enormous and wonderful body; but, as individuals, their ceremonies, their trifling edicts, their jealousy of foreigners, and their tea-cup representations of themselves (which are the only ones popularly known), impress us irresistibly with a fancy that they are a people all toddling, little-eyed, little-footed, little-bearded, little-minded, quaint, overweening, pig-tailed, bald-headed, cone-capped or pagoda-hatted, having childish houses and temples with bells at every corner and story, and shuffling about in blue landscapes, over "nine-inch bridges," with little mysteries of bell-hung whips in their hands—a boat, or a house, or a tree, made of a pattern, being over their heads or underneath them (as the case may happen), and a bird as large as the boat, always having a circular white space to fly in. Such are the Chinese of the tea-cups and the grocers' windows, and partly of their own novels, too, in which everything seems as little as their eyes, little odes, little wine-parties, and a series of little satisfactions. However, it

must be owned, that from these novels one gradually acquires a notion that there is a great deal more good sense and even good poetry among them than one had fancied from the accounts of embassies and the autobiographical paintings on the China-ware; and this is the most probable supposition. An ancient and great nation, as civilised as they, is not likely to be so much behindhand with us in the art of living as our self-complacency leads us to imagine. If their contempt of us amounts to the barbarous, perhaps there is a greater share of barbarism than we suspect in our scorn of them.

At all events, it becomes us to be grateful for their tea. What a curious thing it was, that all of a sudden the remotest nation of the East, otherwise unknown, and foreign to all our habits, should convey to us a domestic custom which changed the face of our morning refreshments; and that, instead of ale and meat, or wine, all the polite part of England should be drinking a Chinese infusion, and setting up earthenware in their houses, painted with preposterous scenery! We shall not speak contemptuously, for our parts, of any such changes in the history of a nation's habits, any more than of the changes of the wind, which now comes from the west, and now from the east, doubtless for some good purpose. It may be noted, that the introduction of tea-drinking followed the diffusion of books among us, and the growth of more sedentary modes of life. The breakfasters upon cold beef and "cool tankards" were an active, horse-riding generation. Tea-drinking times are more indoor, given to reading, and are riders in carriages, or manufacturers at the loom or the steam-engine. It may be taken as an axiom,—the more sedentary, the more tea-drinking. The conjunction is not the best in the world; but it is natural, till something better be found. Tea-drinking is better than dram-drinking: a practice which, if our memory does not deceive us, was creeping in among the politest and even the fairest circles, during the transition from ales to teas. When the late Mr. Hazlitt,* by an effort worthy of him, suddenly left off the stiff glasses of brandy-and-water by which he had been tempted to prop up his disappointments, or rather to loosen his tongue at the pleasant hour of sup-

*William Hazlitt (1778–1830), an important essayist.

per, he took to tea-drinking; and it must be owned, was latterly tempted to make himself as much amends as he could for his loss of excitement in the quantity he allowed himself; but it left his mind free to exercise its powers;—it "kept," as Waller* beautifully says of it,

> The palace of the soul serene;

not, to be sure, the quantity, but the tea itself, compared with the other drink. The prince of tea-drinkers was Dr. Johnson,† one of the most sedentary of men, and the most unhealthy. It is to be feared his quantity suited him still worse; though the cups, of which we hear such multitudinous stories about him, were very small in his time. It was he that wrote, or rather *effused,* the humorous request for tea, in ridicule, of the style of the old ballads (things, be it said without irreverence, which he did not understand so well as "his cups"). The verses were extempore, and addressed to Mrs. Thrale:

> And now, I pray thee, Hetty dear,
>   That thou wilt give to me,
> With cream and sugar soften'd well
>   Another dish of tea.
>
> But hear, alas! this mournful truth,
>   Nor hear it with a frown,—
> Thou canst not make the tea so fast
>   As I can gulp it down.

Now this is among the pleasures of reading and reflecting men over their breakfast, or on any other occasion. The sight of what is a tiresome nothing to others, shall suggest to them a hundred agreeable recollections and speculations. "There is a tea-cup!" a simpleton might cry;—"it holds my tea—that's all." Yes, that's all to you and your poverty-stricken brain; we hope you are rich and

---

*Edmund Waller (1606–1687), English poet.
†Dr. Samuel Johnson (1709–1784), writer, conversationalist, and arbiter of taste whose opinions were chronicled in Boswell's *Life.* His great friend, Mrs. Thrale, was the wife of a wealthy brewer.

prosperous, to make up for it as well as you can. But to the right tea-drinker, the cup, we see, contains not only recollections of eminent brethren of the bohea,* but the whole Chinese nation, with all its history, Lord Macartney† included; nay, for that matter, Ariosto‡ and his beautiful story of Angelica and Medora; for Angelica was a Chinese; and then collaterally come in the Chinese neighbours and conquerors from Tartary, with Chaucer's

> Story of Cambuscan bold,

and the travels of Marco Polo and others, and the Jesuit missionaries, and the Japanese with our friend Golownin,§ and the Loo Choo people, and Confucius, whom Voltaire (to show his learning) delights to call by his proper native appellation of Kong-footsee (reminding us of Congo tea); and then we have the *Chinese Tales*, and Goldsmith's‖ *Citizen of the World*, and Goldsmith brings you back to Johnson again and the tea-drinkings of old times; and then we have the *Rape of the Lock* before us, with Belinda at breakfast,¶ and Lady Wortley Montague's tea-table eclogue,** and the domestic pictures in the *Tatler* and *Spectator*,†† with the passions existing in those times for china-ware, and Horace Walpole,‡‡ who was an old woman in that respect; and, in short, a thousand other memories, grave and gay, poetical and prosaical, all ready to wait upon anybody who chooses to read books, like spirits at the command of the book-readers of old, who, for the advantages they had over the rest of the world, got the title of Magicians.

* Hills in china where black tea was grown, which in the eighteenth century was regarded as the best but now thought to be inferior. The word is Chinese for "white tip" and refers to the leaves.
† George Macartney (1737–1806), the first British emissary to Peking sent to China on a trade mission that failed.
‡ Ludovico Ariosto (1474–1533), Italian poet known for his epic *Orlando Furioso*.
§ Vasily Mikhaylovich Golownin: (1776–1831), Russian naval officer sent to chart Alaska and captured by the Japanese in 1810. His *Narrative of My Captivity in Japan* (1810–1813) was published in 1816 and stimulated interest in the Orient.
‖ Oliver Goldsmith (1730–1774), essayist and poet.
¶ Alexander Pope's 1714 comic poem about a coquette, Belinda, whose would-be suitor cuts off a lock of her hair in triumph at a card party. The current fashions of tea, coffee, and chocolate drinking are described in the poem.
** Lady Mary Wortley Montague (1689–1762).
†† Early eighteenth-century magazines by Joseph Addison and Richard Steele that chronicled London life and fashions.
‡‡ Horace Walpole (1717–1797), a distinguished antiquarian and writer.

Yea, pleasant and rich is thy sight, little tea-cup (large, though, at breakfast), round, smooth, and coloured;—composed of delicate earth—like the earth, producing flowers, and birds, and men; and containing within thee thy Lilliputian ocean, which we, after sending our fancy sailing over it, past islands of foam called "sixpences," and mysterious bubbles from below, will, giant-like, engulf;—

But hold—there's a fly in.

Now, why could not this inconsiderate monster of the air be content with the whole space of the heavens round about him, but he must needs plunge into this scalding pool? Did he scent the sugar? or was it a fascination of terror from the heat? "Hadst thou my three kingdoms to range in," said James the First to a fly, "and yet must needs get into my eye?" It was a good-natured speech, and a natural. It shows that the monarch did his best to get the fly out again; at least we hope so; and therefore we follow the royal example in extricating the little winged wretch, who has struggled hard with his unavailing pinions, and becomes drenched and lax with the soaking. He is on the dry, clean cloth. Is he dead? No:—the tea was not so hot as we supposed it:—see, he gives a heave of himself forward; then endeavours to drag a leg up, then another, then stops, and sinks down, saturated and overborne with wateriness; and assuredly, from the inmost soul of him, he sighs (if flies sigh—which we think they must do sometimes, after attempting in vain, for half an hour, to get through a pane of glass). However, his sigh is as much mixed with joy as fright and astonishment and a horrible hot bath can let it be; and the heat has not been too much for him; a similar case would have been worse for one of us with our fleshy bodies;—for see! after dragging himself along the dry cloth, he is fairly on his legs; he smoothes himself, like a cat, first on one side, then the other, only with his legs instead of his tongue; then rubs the legs together, partly to disengage them of their burthen, and partly as if he congratulated himself on his escape; and now, finally, opening his wings (beautiful privilege! for all wings, except the bat's, seem beautiful, and a privilege, and fit for envy), he is off again into the air, as if nothing had happened.

He may forget it, being an inconsiderate and giddy fly; but it is to us, be it remembered by our conscience, that he owes all which he is hereafter to enjoy. His suctions of sugar, his flights, his dances on the window, his children, yea, the whole House of Fly, as far as it depends on him their ancestor, will be owing to us. We have been his providence, his guardian angel, the invisible being that rescued him without his knowing it. What shall we add, reader? Wilt thou laugh, or look placid and content—humble, and yet in some sort proud withal, and not consider it as an unbecoming meeting of ideas in these our most mixed and reflective papers— if we argue from rescued flies to rescued human beings, and take occasion to hope, that in the midst of the struggling endeavours of such of us as have to wrestle with fault or misfortune, invisible pity may look down with a helping eye upon ourselves, and that what it is humane to do in the man, it is divine to do in that which made humanity.

🍸

DAISETZ T. SUZUKI

(JAPANESE, 1870–1966)

From *Zen and Japanese Culture*

THE TEAMAN AND THE RUFFIAN

What follows is the story of a teaman who had to assume the role of a swordsman and fight with a ruffian. The teaman generally does not know anything about swordplay and cannot be a match in any sense of the word for anybody who carries a sword. His is a peaceful profession. The story gives us an idea of what a man can do with a sword even when he has never had any technical training, if only his mind is made up to go through the business at the risk of his life. Here is another illustration demonstrating the value of resoulte-mindedness leading up to the transcendence of life and death.

Toward the end of the seventeenth century, Lord Yama-no-uchi, of the province of Tosa, wanted to take his teamaster along with him on his official trip to Yedo, the seat of the Tokugawa Shogunate. The teamaster was not inclined to accompany him, for in the first place he was not of the samurai rank and knew that Yedo was not a quiet and congenial place like Tosa, where he was well known and had many good friends. In Yedo he would most likely get into trouble with ruffians, resulting not only in his own disgrace but in his lord's. The trip would be a most risky adventure, and he had no desire to undertake it.

The lord, however, was insistent and would not listen to the remonstrance of the teamaster; for this man was really great in his profession, and it was probable that the lord harbored the secret desire to show him off among his friends and colleagues. Not able to resist further the lord's earnest request, which was in fact a command, the master put off his teaman's garment and dressed himself as one of the samurai, carrying two swords.

While staying in Yedo, the teamaster was mostly confined in his lord's house. One day the lord gave him permission to go out and do some sight-seeing. Attired as a samurai, he visited Uyeno by the Shinobazu pond, where he espied an evil-looking samurai resting on a stone. He did not like the looks of this man. But finding no way to avoid him, the teaman went on. The man politely addressed him: "As I observe, you are a samurai of Tosa, and I should consider it a great honor if you permit me to try my skill in swordplay with you."

The teaman of Tosa from the beginning of his trip had been apprehensive of such an encounter. Now, standing face to face with a *rōnin* * of the worst kind, he did not know what to do. But he answered honestly: "I am not a regular samurai, though so dressed; I am a teamaster, and as to the art of swordplay I am not at all prepared to be your opponent." But as the real motive of the *rōnin* was to extort money from the victim, of whose weakness he was now fully convinced, he pressed the idea even more strongly on the teaman of Tosa.

Finding it impossible to escape the evil-designing *rōnin,* the teaman made up his mind to fall under the enemy's sword. But he did not wish to die an ignominious death that would surely reflect

* Self-employed samurai.

on the honor of his lord of Tosa. Suddenly he remembered that a few minutes before he had passed by a swordsman's training school near Uyeno park, and he thought he would go and ask the master about the proper use of the sword on such occasions and also as to how he should honorably meet an inevitable death. He said to the *rōnin*, "If you insist so much, we will try our skill in swordsmanship. But as I am now on my master's errand, I must make my report first. It will take some time before I come back to meet you here. You must give me that much time."

The *rōnin* agreed. So the teaman hastened to the training school referred to before and made a most urgent request to see the master. The gatekeeper was somewhat reluctant to acquiesce because the visitor carried no introductory letter. But when he noticed the seriousness of the man's desire, which was betrayed in his every word and in his every movement, he decided to take him to the master.

The master quietly listened to the teaman, who told him the whole story and most earnestly expressed his wish to die as befitted a samurai. The swordsman said, "The pupils who come to me invariably want to know how to use the sword, and not how to die. You are really a unique example. But before I teach you the art of dying, kindly serve me a cup of tea, as you say you are a teaman." The teaman of Tosa was only too glad to make tea for him, because this was in all likelihood the last chance for him to practice his art of tea to his heart's content. The swordsman closely watched the teaman as the latter was engaged in the performance of the art. Forgetting all about his approaching tragedy, the teaman serenely proceeded to prepare tea. He went through all the stages of the art as if this were the only business that concerned him most seriously under the sun at that very moment. The swordsman was deeply impressed with the teaman's concentrated state of mind, from which all the superficial stirrings of ordinary consciousness were swept away. He struck his own knee, a sign of hearty approval, and exclaimed, "There you are! No need for you to learn the art of death! The state of mind in which you are now is enough for you to cope with any swordsman. When you see your *rōnin* outcast, go on this way: First, think you are going to serve tea for a guest. Courteously salute him, apologizing for the delay, and tell him that you are now ready for the contest. Take off your *haori* [outer coat], fold it up carefully, and then put your fan on it just as you do when you are at work. Now bind your head with the *tenugui* [corresponding to a towel], tie your sleeves up with the string, and gather up your *hakama* [skirt]. You are now prepared for the business that is to start immediately. Draw your sword, lift it high up over your head, in full readiness to strike down the opponent, and, closing your eyes, collect your thoughts for a combat. When you hear him give a yell, strike him with your sword. It will probably end in a mutual slaying." The teaman thanked the master for his instructions and went back to the place where he had promised to meet the combatant.

He scrupulously followed the advice given by the swordmaster with the same attitude of mind as when he was serving tea for his friends. When, boldly standing before the *rōnin*, he raised his sword, the *rōnin* saw an altogether different personality before him. He had no chance to give a yell, for he did not know where and how to attack the teaman, who now appeared to him as an embodiment of fearlessness, that is, of the Unconscious. Instead of advancing toward the opponent, the *rōnin* retreated step by step, finally crying, "I'm done, I'm done!" And, throwing up his sword, he prostrated himself on the ground and pitifully asked the teaman's pardon for his rude request, and then he hurriedly left the field.

♉

MARK VAN DOREN

(A M E R I C A N , 1 8 9 4 – 1 9 7 2)

Tea

The time for it, I think,
Is marked upon the wall
Between my mind and yours—
Oh, but there is one,

Of air and isinglass,*
And it is made to fall
As time is made to pass.

Time and tea together
Are such a magic thing
As emperors once dreamed of;
And empresses, left lonely
At darkening of day,
Clapped hands as if to frighten
The woe of the world away.

♮

H. H. MUNRO [SAKI]

(E N G L I S H , 1 8 7 0 – 1 9 1 6)

Tea

James Cushat-Prinkly was a young man who had always had a settled conviction that one of these days he would marry; up to the age of thirty-four he had done nothing to justify that conviction. He liked and admired a great many women collectively and dispassionately without singling out one for especial matrimonial consideration, just as one might admire the Alps without feeling that one wanted any particular peak as one's own private property. His lack of initiative in this matter aroused a certain amount of impatience among the sentimentally minded women-folk of his home circle; his mother, his sisters, an aunt-in-residence, and two or three intimate matronly friends regarded his dilatory approach to the married state with a disapproval that was far from being inarticulate. His most innocent flirtations were watched with the straining eagerness which a group of unexercised terriers concen-

*Thin membrane.

trates on the slightest movements of a human being who may be reasonably considered likely to take them for a walk. No decent-souled mortal can long resist the pleading of several pairs of walk-beseeching dog-eyes; James Cushat-Prinkly was not sufficiently obstinate or indifferent to home influences to disregard the obviously expressed wish of his family that he should become enamoured of some nice marriageable girl, and when his Uncle Jules departed this life and bequeathed him a comfortable little legacy it really seemed the correct thing to do to set about discovering some one to share it with him. The process of discovery was carried on more by the force of suggestion and the weight of public opinion than by any initiative of his own; a clear working majority of his female relatives and the aforesaid matronly friends had pitched on Joan Sebastable as the most suitable young woman in his range of acquaintance to whom he might propose marriage, and James became gradually accustomed to the idea that he and Joan would go together through the prescribed stages of congratulations, present-receiving, Norwegian or Mediterranean hotels, and eventual domesticity. It was necessary, however, to ask the lady what she thought about the matter; the family had so far conducted and directed the flirtation with ability and discretion, but the actual proposal would have to be an individual effort.

Cushat-Prinkly walked across the Park towards the Sebastable residence in a frame of mind that was moderately complacent. As the thing was going to be done he was glad to feel that he was going to get it settled and off his mind that afternoon. Proposing marriage, even to a nice girl like Joan, was a rather irksome business, but one could not have a honeymoon in Minorca and a subsequent life of married happiness without such preliminary. He wondered what Minorca was really like as a place to stop in; in his mind's eye it was an island in perpetual half-mourning, with black or white Minorca hens running all over it. Probably it would not be a bit like that when one came to examine it. People who had been in Russia had told him that they did not remember having seen any Muscovy ducks there, so it was possible that there would be no Minorca fowls on the island.

His Mediterranean musings were interrupted by the sound of a clock striking the half-hour. Half-past four. A frown of dissatisfaction settled on his face. He would arrive at the Sebastable mansion just at the hour of afternoon tea. Joan would be seated at a low table, spread with an array of silver kettles and cream-jugs and delicate porcelain teacups, behind which her voice would tinkle pleasantly in a series of little friendly questions about weak or strong tea, how much, if any, sugar, milk, cream, and so forth. "Is it one lump? I forgot. You do take milk, don't you? Would you like some more hot water, if it's too strong?"

Cushat-Prinkly had read of such things in scores of novels, and hundreds of actual experiences had told him that they were true to life. Thousands of women, at this solemn afternoon hour, were sitting behind dainty porcelain and silver fittings, with their voices tinkling pleasantly in a cascade of solicitous little questions. Cushat-Prinkly detested the whole system of afternoon tea. According to his theory of life a woman should lie on a divan or couch, talking with incomparable charm or looking unutterable thoughts, or merely silent as a thing to be looked on, and from behind a silken curtain a small Nubian page should silently bring in a tray with cups and dainties, to be accepted silently, as a matter of course, without drawn-out chatter about cream and sugar and hot water. If one's soul was really enslaved at one's mistress's feet, how could one talk coherently about weakened tea? Cushat-Prinkly had never expounded his views on the subject to his mother; all her life she had been accustomed to tinkle pleasantly at tea-time behind dainty porcelain and silver, and if he had spoken to her about divans and Nubian pages she would have urged him to take a week's holiday at the seaside. Now, as he passed through a tangle of small streets that led indirectly to the elegant Mayfair terrace for which he was bound, a horror at the idea of confronting Joan Sebastable at her tea-table seized on him. A momentary deliverance presented itself; on one floor of a narrow little house at the noisier end of Esquimault Street lived Rhoda Ellam, a sort of remote cousin, who made a living by creating hats out of costly materials. The hats really looked as if they had come from Paris; the cheques she got for them unfortunately never looked as if they were going to Paris. However, Rhoda appeared to find life amusing and to have a fairly good time in spite of her straitened circumstances. Cushat-Prinkly decided to climb up to her floor and defer by half-an-hour or so the important business which lay before him; by spinning out his visit he could contrive to reach the Sebastable mansion after the last vestiges of dainty porcelain had been cleared away.

Rhoda welcomed him into a room that seemed to do duty as workshop, sitting-room, and kitchen combined, and to be wonderfully clean and comfortable at the same time.

"I'm having a picnic meal," she announced. "There's caviare in that jar at your elbow. Begin on that brown bread-and-butter while I cut some more. Find yourself a cup; the teapot is behind you. Now tell me about hundreds of things."

She made no other allusion to food, but talked amusingly and made her visitor talk amusingly too. At the same time she cut the bread-and-butter with a masterly skill and produced red pepper and sliced lemon, where so many women would merely have produced reasons and regrets for not having any. Cushat-Prinkly found that he was enjoying an excellent tea without having to answer as many questions about it as a Minister for Agriculture might be called on to reply to during an outbreak of cattle plague.

"And now tell me why you have come to see me," said Rhoda suddenly. "You arouse not merely my curiosity but my business instincts. I hope you've come about hats. I heard that you had come into a legacy the other day, and, of course, it struck me that it would be a beautiful and desirable thing for you to celebrate the event by buying brilliantly expensive hats for all your sisters. They may not have said anything about it, but I feel sure the same idea has occurred to them. Of course, with Goodwood on us, I am rather rushed just now, but in my business we're accustomed to that; we live in a series of rushes—like the infant Moses."

"I didn't come about hats," said her visitor. "In fact, I don't think I really came about anything. I was passing and I just thought I'd look in and see you. Since I've been sitting talking to you, however, a rather important idea has occurred to me. If you'll forget Goodwood for a moment and listen to me, I'll tell you what it is."

Some forty minutes later James Cushat-Prinkly returned to the bosom of his family, bearing an important piece of news.

"I'm engaged to be married," he announced.

A rapturous outbreak of congratulation and self-applause broke out.

"Ah, we knew! We saw it coming! We foretold it weeks ago!"

"I'll bet you didn't," said Cushat-Prinkly. "If any one had told me at lunch-time today that I was going to ask Rhoda Ellam to marry me and that she was going to accept me, I would have laughed at the idea."

The romantic suddenness of the affair in some measure compensated James's women-folk for the ruthless negation of all their patient effort and skilled diplomacy. It was rather trying to have to deflect their enthusiasm at a moment's notice from Joan Sebastable to Rhoda Ellam; but, after all, it was James's wife who was in question, and his tastes had some claim to be considered.

On a September afternoon of the same year, after the honeymoon in Minorca had ended, Cushat-Prinkly came into the drawing-room of his new house in Granchester Square. Rhoda was seated at a low table, behind a service of dainty porcelain and gleaming silver. There was a pleasant tinkling note in her voice as she handed him a cup.

"You like it weaker than that, don't you? Shall I put some more hot water to it? No?"

THOMAS HARDY

(1840–1928)

At Tea

The kettle descants in a cosy drone,
And the young wife looks in her husband's face,
And then at her guest's, and shows in her own
Her sense that she fills an envied place;
And the visiting lady is all abloom,
And says there was never so sweet a room.

And the happy young housewife does not know
That the woman beside her was first his choice,

Till the fates ordained it could not be so. . . .
Betraying nothing in look or voice
The guest sits smiling and sips her tea,
And he throws her a stray glance yearningly.

SAMUEL TAYLOR COLERIDGE

(E N G L I S H , 1 7 7 2 – 1 8 3 4)

Monody on a Tea-Kettle

O Muse who sangest late another's pain,
To griefs domestic turn thy coal-black steed!
With slowest steps thy funeral steed must go,
Nodding his head in all the pomp of woe:
Wide scatter round each dark and deadly weed,
And let the melancholy dirge complain,
(Whilst Bats shall shriek and Dogs shall howling run)
The tea-kettle is spoilt and Coleridge is undone!

Your cheerful songs, ye unseen crickets, cease!
Let songs of grief your alter'd minds engage!
For he who sang responsive to your lay,
What time the joyous bubbles 'gan to play,
The *sooty swain* has felt the fire's fierce rage;—
Yes, he is gone, and all my woes increase;
I heard the water issuing from the wound—
No more the Tea shall pour its fragrant steams around!

O Goddess best belov'd! Delightful Tea!
With thee compar'd what yields the madd'ning Vine?
Sweet power! who know'st to spread the calm delight,
And the pure joy prolong to midmost night!
Ah! must I all thy varied sweets resign?

Enfolded close in grief thy form I see;
No more wilt thou extend thy willing arms,
Receive the *fervent Jove,* and yield him all thy charms!

How sink the mighty low by Fate opprest!—
Perhaps, O Kettle! thou by scornful toe
Rude urg'd t' ignoble place with plaintive din,
May'st rust obscure midst heaps of vulgar tin;—
As if no joy had ever seiz'd my breast
When from thy spout the streams did arching fly,—
As if, infus'd, thou ne'er hadst known t' inspire
All the warm raptures of poetic fire!

But hark! or do I fancy the glad voice—
"What tho' the swain did wondrous Charms disclose—
(Not such did Memnon's* sister sable drest)
Take these bright arms with royal face imprest,
A better Kettle shall thy soul rejoice,
And with Oblivion's wings o'erspread thy woes!"
Thus Fairy Hope can soothe distress and toil;
On empty Trivets she bids fancied Kettles boil!

♈

MARY LAVIN

(IRISH, BORN AMERICA, 1912–)

A Cup of Tea

"She'll take a cup of tea, no matter how late it is when she gets here. You can leave the kettle at the back of the stove where it will keep the heat without spilling over and putting out the fire."

"All right, ma'am," said the servant girl, and she threw a baleful

---

*Child of Tithonus and Eos.

glance at a picture of Sophy for whom all this trouble was being taken, and who had been expected since early morning and, would, as likely as not, arrive in the middle of the night, and get them all up out of their beds. The young servant's back was nearly broken from all the work that had been done in the last few days. They had scrubbed every floor in the house, cleaned all the windows, and they had gone as far as waxing Sophy's room twice in one day because Sophy's mother had marked it so much with tracks of her feet going in and out with clean curtains and extra pillows, bunches of flowers and hot jars.

"She needs a holiday after all the hard work she has done," said her mother as she pushed the kettle further back on the stove, and then changed her mind and left it where it had been.

"I hope she'll pass her examination," said the servant, raising up from the floor in the hope of gaining a moment's rest by introducing a topic upon which her mistress always showed weakness and garrulity.

"Her examination!"

It was almost as if Sophy's mother had forgotten that Sophy had done any examination at all, so absorbed was she in the thought that her only daughter was coming home again after a three months' absence.

"Oh, her examination!" she repeated, carelessly as her eyes ran over the tray that was already set for Sophy's breakfast in bed next morning.

"Why, of course she'll pass. I expect she'll do very well, and get very good marks."

"She's very clever, isn't she?" said the young girl, in a tone of voice in which she was careful to mingle mixed ingredients of envy and flattery, as a good farmer mingles his grasses with cocksfoot and ratstail and clover when he lays down a field.

"Oh, she has a good brain, I suppose," said Sophy's mother, in an offhand manner that so far from deceiving the servant emboldened her to sit back on her hunkers and rest her limbs more freely.

"Why wouldn't she have brains!" said the young girl, "I suppose she takes after her father!" And jerking her finger and thumb she was about to point in dumb show at the ceiling, for over their head was the study in which Sophy's father was buried in books and lost in the fumes of tobacco smoke. He was an amateur entomologist.

But Sophy's mother turned around sharply.

"What time do you expect to get that floor scoured? Do you expect to be at it all night, or why are you sitting back and taking your ease like that?"

The girl grabbed the scouring brush again and began to rub it hastily on the big bar of soap that she had held indifferently in her hand while she had taken her ease.

Except for the noise of the rough brush going over the flagged floor there was no sound then for a few minutes, but it was clear from the expression upon her face that Sophy's mother was still brooding on what the servant had said.

"There's a great deal of difference between a hobby and a degree from a University!" she said at last, "and Sophy has done one of the hardest degrees in the University. There were only two girls in the Political Economy class, and only one in the class for the Study of International Relations." She smiled. "Her father spends a lot of his time up there in his study with his books and his collection-cases, but I hardly think that he'd make much out of Sophy's books! Did you ever see them? I never saw such books. So heavy! And figures and diagrams on every second page."

During this speech the mother's face relaxed somewhat and assumed an expression of satisfaction such as it had worn after Sophy's floor had been waxed the second time. "Of course, my husband is a great authority on insects," she said, grudgingly, as one who having very little respect for the position of another is nevertheless aware that upon the respect in which it is held by others his own importance depends. She was going to say something else, but just then the young servant ran to the window and began to pull off her apron and struggle to get it over her head.

"She's here! She's here!" cried the girl excitedly. And then, before the mother was into the living room and only half-way across the floor to the hallway, Sophy was running across the room to meet her, throwing her bag and scarf to either side of her, and putting out her arms to hug her mother with such a rattle of bracelets and metal buttons that although the cheeks that pressed against hers were soft and firm as plums and peaches her mother was reminded for a moment of Sophy's father and the way his watch-chain used to rattle against the buttons on his waistcoat long ago when he gave her a salutation. That was before he took to wearing smoking jackets with buttons of braid or velvet, and gave up bothering to carry around a watch of any kind. But the impression only lasted a moment.

"Stand back from me," she cried. "Let me look at you! You look thin. I hope you're not trying to slim yourself. How did the car drive up without my hearing it? Did you have a tiresome journey? What time did you leave? I hope you stopped on the road for a rest!"

The mother's eyes flew from Sophy's face to her waist, from her waist to her hair, and from her hair to her hands with that confused, distracted, and haphazard curiosity of love that is so different from the steady gaze of a stranger.

"How did you get on at your examination?" she asked, and then before Sophy had time to answer even the last of the questions, her mother threw up her hands. "Hold up your head! Don't stir!" she said, and she moved back as one moves back from a canvas at a picture gallery, the better to see it. "You're too thin," she said. "It doesn't suit you. It throws up your likeness to your father. I often heard people drawing attention to it, but I could never see it myself. But I see it now! I'd swear I was looking at him, except for his moustaches. How odd I never saw it before. Sometimes, I thought there was a slight resemblance in the way you turned your head, but only at certain times, if you were upset or annoyed about something. I can't say it was ever anything definite. It was nothing at all like the way you resemble my own sisters in your walk and the way you hold yourself. But I suppose it's natural that you should resemble him in something. It's a good thing you didn't inherit his disposition!"

As she made the last remark she felt Sophy draw away her hand, which she had still kept clasped in hers while leaning back to scrutinize her. Instantly she regretted having let her tongue run away with her, but Sophy's face gave no sign of annoyance. She had released her hands, certainly, but she had smiled as she moved away and began to gather up her scarf, her bag and her gloves. Still her mother regretted the remark.

"How is he?" said the girl, and then suddenly she sank down into an armchair in case her mother might think for a moment that she was impatient to go up to him.

"You'll have to go up and see him. He musn't have heard the car!" said her mother grudgingly, and then before she could stop herself she had given in to her spite again. "I don't know why he didn't hear the car! His study is to the side of the house, and I was in the kitchen, away at the back!" But almost at once, she made an effort to undo the harm she had done. "You had better run up and see him before you take off your hat and coat," she said, and then she urged herself to be even more pressing. "Please, dear!" she said, briskly. "Don't sit down until you have gone up to him, if only for a minute. You need only say a few words; just to please him. There's no need to stay. Just a few words! Don't sit down till you do it. If you sit down you won't feel like getting up." She walked over and taking her daughter's arm urged her to stand up by pressing her arm. "I'll have your cases brought up to your room while you're gone, and when you come down we'll have a nice chat. I want to hear all you have to tell me. But not until you come down again!"

Sophy began to get up from the armchair slowly. She pouted and put on an air of unwillingness, feeling compelled by an impulse of pity and compassion to play up to the pretence of her own reluctance with which her mother was deluding herself.

"Hurry, dear! You'll be glad you went. Just tell him you got here safely. He'd feel so hurt if you didn't go up."

"Why didn't he come down?" said Sophy, with a good appearance of resentment.

"He's coming to the final chapters of his work on those disgusting beetles he has all over the house," said her mother, and she shuddered. "He comes down for his meals because I insist, but he doesn't come down at all in the evenings. I see his light burning till all hours in the night. I say the book will want to bring in queer profits before it pays for all the lamp-oil he's burned in the last five years! Although who's going to buy it is more than I can make out unless there are more than I think of dried-up unnatural people like himself, that see more value in a bee's foot than they see in the company of the human beings around them!"

Again she recollected herself quickly and rushed on without drawing a breath. "But go up," she cried. "Go to the door for a minute, anyway. There's no need to stay."

But as Sophy went out of the room dragging her feet, the mother knew in her heart that the girl would not come down again until she had to be called, and perhaps called more than once. Even as she listened she heard the footsteps getting faster. They stopped dragging and soon broke into a run, taking the stairs two at a time. Suddenly she thought of the lassitude and disinclination that Sophy had shown about going upstairs, although she had been as fresh as a daisy when she first came in, and seemed to freshen up quickly enough when she got to the stairs. Had she too been pretending? And why? Her mother regretted her own polite pretence, and almost ran out into the hallway to shout up the stairs.

"Go on! Don't let me keep you. Go on. Run up to your father. Tell him all your news. Don't think of me. Don't tell me anything at all."

But she shut her mouth tightly and went out to tell the maid that she could go to bed, there was nothing more to be done.

The maid was worn out. She sat on a chair in front of the range with her legs apart and her head bent as she tried to take a splinter out of her thumb by sucking it.

"What is the matter? Is your finger sore? Put a poultice of bread and water on it and leave it on till morning." She looked around the kitchen. "You can go to bed now," she said, and then she remembered the bags. "You can carry up the baggage to her room before you go," she said, and then she went over to the pantry door. "Did you leave out a jug of milk?" she asked, "and a bowl of sugar? I forgot to ask her if she'd have a cup of tea, but I'm sure she'll want one."

The girl rose up unsteadily to her feet and went into the pantry. She came out with a large jug of milk, but her face had a dubious expression, and when she came into the light of the kitchen she put the jug to her nose and smelled it, making a grimace as she did so.

"Don't do that," shouted her mistress, nearly causing the girl to drop the jug. "Where did you pick up that disgusting habit?"

"I only wanted to see if the milk was sour. How could I find out without smelling it?" said the girl, not fully aware of the error of what she had done, but painfully aware of the error of having been seen doing it.

"The milk couldn't be sour. It's the afternoon's milk. Give it to me!"

"All the same," said the girl, and she handed over the jug, looking with interest to see in what superior way its condition would be tested by her mistress.

Sophy's mother took the jug and for a minute she seemed to hesitate, then she began to move it deliberately round and round in concentric circles a long way from her nose, but presumably within the orbit of her olefactory organs, for as she did so she sniffed two or three times. But the presumption was wrong.

"There is no need to stick your nose right into the jug," she said, then, feeling a need to modify her former remarks on the subject, as she raised the jug and brought it slightly nearer to her nose, before she circled it round once more. This time her nostrils dilated slightly. The young girl tossed her head and took out a hair-slide rather unnecessarily since she fastened it in her hair again a minute later in the exact same place where it had been before. "Don't put your hands to your hair in the kitchen," said her lady absentmindedly, but she was looking around the kitchen. "Where is the small saucepan?" she said. "Oh, there it is. Rinse it out with clean water for me. And then you can go to bed. This milk is perfectly fresh, but it will do it no harm to give it a boil. That will make sure of it for the morning. It will be cool before Sophy comes downstairs, I'm sure."

The milk was well cooled when Sophy came down. Her mother had been up and down to her own room two or three times and had at last undressed and put on her slippers and dressing-gown. She walked heavily each time she passed her husband's study, but the talk and laughing inside may have prevented her footsteps from being heard. When she passed the boxroom under the stairs she heard the young servant snoring and groaning in her sleep.

"Oh Mother! I kept you up," cried Sophy when she came down at last and saw the slippers and dressing-gown. "Why did you wait up? Why didn't you call me?" And then, as if answering both of her own questions, she summed up the situation. "You should have gone to bed," she said.

"I thought you might like some tea," said her mother. "I kept the fire in. It was nearly out, but I was just in time to throw on a few logs."

"Oh good!" said Sophy.

Her mother's spirits that had sunk in the last hour began to rise again.

"I have the tray set," she said, "and I'll have a cup with you. I never take tea at this hour because it keeps me awake but I'll be awake tonight anyway, with excitement!"

Her spirits rose higher. She forgot her hurt at the hour Sophy had spent upstairs.

"I want to hear everything that happened since you went away. Begin at the beginning. Were you wet when you reached the station going back last time? I worried all night thinking of the long journey you had sitting in that cold railway carriage. They should have proper heating apparatus. It's disgraceful. Did you meet anyone going up in the train? Were you back on the right day? Were there any new people in your class?"

"I'm sure I told you all that in a letter," said Sophy, taken aback at her mother's accurate memory of a day she had difficulty in recalling.

"I could hear it a dozen times," said her mother. "Tell me everything!"

"Oh, there's nothing unusual to tell," said Sophy taking up a biscuit and beginning to chew it.

"You found plenty to tell your father it seems, judging by the length of time you stayed up with him," said her mother irritably.

Sophy looked up. "Oh, we were only talking," she said vaguely, for she could not, in fact, remember much of the lazy banter that she and her father exchanged when they met. "There's nothing much to tell. The exams were not too bad. The results will be in the paper." She paused and tried to remember something else to tell. "The second paper on the first day was a bit hard, but I think I'll pull through all right."

"I can hardly believe it's your final examination," said her mother. "How time flies! I lie awake nights wondering if we did right in sending you to the University. It puts such a strain on a girl. I hope you'll stay home for a while, even after the results come out.

Indeed, if you like I'll speak to your father about letting you stay home entirely. It's very nice to have a degree and feel independent, but there's no need to carry things too far and wear yourself out working when your father himself makes no attempts to add to his income, although he could earn more than any man in the town if he wasn't so odd!"

"You have the wrong idea, Mother. I had a definite idea in taking out my degree."

The mother looked at her, at her glossy silk knees and her nice straight hair.

"Have you some plans?" she asked excitedly. "I'm so happy to have you home. Tell me your plans. I want to hear them. I want to help."

She remembered the way she went into her own mother's room years ago to sit in an armchair and brush her hair when she came back from a visit to a relation, telling her mother all about the dances to which she had gone, the partners with whom she had danced, about the underlinen her cousins had worn, the needlework they were doing, and even arranging her hair in the manner in which they had worn theirs so that nothing would be unknown to those at home of the way her time had been spent during her absence from them.

"One thing is certain," she said suddenly, speaking out of a dream. "You have my mother's hands. I never saw anything so remarkable. She would have been so pleased to know it, but I never thought of telling her." She sighed and her faint sigh seemed to fill the air with ghostly memories and regrets, and Sophy moved in her chair, and stared down at her hands self-consciously, but seeing a loosened piece of cuticle on the forefinger of her left hand she became absorbed at once in trying to tear it off with the finger and thumb of her right hand, which she crooked to form a pincers. She gave two or three sharp tugs to the loose piece of skin but failed to tear it off each time. "Don't do that," said her mother. "It will be sore all night. It will keep you awake."

"I can't leave it alone," said Sophy, tugging at it again.

"You can. Forget about it. Stop looking at it. Tell me more about your examination. Tell me about your friends and you'll forget all about it, and it will knit back into the skin."

But Sophy could not forget the dead cuticle. It irritated her, and she kept at it. The mother spoke again in a dreamy voice.

"Poor mother's hands were all blackened and hardened from work, but she wouldn't have grudged you your easy life. She would have been the first to congratulate you on your independence in going to the University. How it would have amused her to see your hands so much like her own. I wish I had thought of telling her. It's only when people are gone from us that we remember all the little things we could have done to please them!"

"Ah!" said Sophy. She had torn off the cuticle and she put her finger up to her mouth to ease the sting of pain. Then she looked at the tray. "How about the tea?" she said.

"The kettle is at the back of the stove," said her mother. "It will be almost boiling. I'll go out and get it. If we poke up the fire here it will bring the water to the boil in a minute."

"Is there any use bringing up tea to father?" said Sophy when her mother came back into the room with the kettle.

"You can if you like," said the woman bending down to settle the kettle on the crackling logs in the grate. "I know the reception I'd get if I interrupted him at his work! Anyone would imagine it was for my own good I went up to him, to hear the way he turned on me upon the few occasions I was unwise enough to enquire if he'd like anything before I went to bed!"

"Oh, I'm so hungry! I'm so thirsty!" said Sophy, breaking in upon her mother urgently. "All the way down in the train I was thinking of those dear old cups with the impossible birds on them." She held up one of the frail cups and stared at its hand-painted rim. But her mother understood the interruption and flushed.

"If that's the case it's a wonder you stayed upstairs so long," she said bitterly, and once again, incongruously, she remembered the confidences that used to be exchanged between her own mother and herself long ago. It would have been such a relief, just once, to tell Sophy of some of the unpleasant things she had had to put up with while she was away. But Sophy was talking about the stupid old cups as if she wanted to stave off any chance of confidences. She was holding the cup in front of her with both hands and staring at it with burning cheeks and bright eyes. How could she be so excited about an old cup that she had seen in the china closet ever since she was able to toddle!

"Is it meant to be a real bird?" Sophy cried. "Is it a peacock? Is it a nightingale? Whoever saw such feathers! Whoever saw such colours!" Her voice was feverish and she asked question after question, laughing nervously at the same time. But in her heart she was trying to form some sort of a prayer. "Dear God, keep her from complaining about Father! I want to comfort her. I want to be understanding. I want to make her happy on my first night home. But I can't see things from her point of view. I can't see things from her point of view."

"Give me the cup," said her mother. "The tea is ready." She caught the small cup almost roughly. She disliked it with a sudden passion and could have thrown it on the floor only it was belonging to a set. Instead, she began to talk as normally as she could. "I'm looking forward to a cup of tea myself," she said. "Indeed, I'd take a cup every night if I had someone to take one with me. I used to love a cup at this hour. I don't know why I gave it up. You give up a great many things, one after another, when you are alone. Of course, I don't want company. I have my reading! I have my knitting! I have a big basket of socks to darn. It never seems to be empty!"

Poor mother. Sophy looked at her mother's face as it was bent over the fire and for a moment she envied the fine structure of bone that made the face so clear and attractive in spite of age, and she looked at the grey hair that was still so full of life that it sprang into curling tendrils whenever it escaped from the combs that held it. She thought of her own large pale face and straight hair that gave her a resemblance to her father that in spite of her mother's difficulty in seeing it, she herself knew to be evident.

"The kettle is beginning to sing," said her mother, just then. "It's a good thing someone in the house has the heart to sing, as I often say to the maid, when your father has been particularly trying."

Sophy looked around quickly for something with which she could attract her mother's attention away from the abyss over which she trembled again. Her eyes fell on the old album, that her mother had taken from her old home, when she had left to get married. It was covered in faded blue velvet, with a large silver clasp.

"Did you paste in the picture of my class that I sent you at mid-term?" she asked, and she leaned forward and took the album on her knee. It fell open at the page upon which she had gazed so frequently as a child, and which she had probably strained from the binding by the many times she leaned her elbows on the book to gaze better at the familiar picture. It was a photograph of her grandmother sitting in a wicker chair against a photographer's scenic screen of palms and clouds and trellis balustrades, and around her, grouped in formal postures, were her daughters, Sophy's mother, and Sophy's mother's sisters. And they all wore long white dresses with long white sleeves and high necks and white hats that tilted under the weight of floppy silk roses. They all wore gold brooches, and gold bracelets, and they all had great masses of nut-brown hair. And they sat with their arms entwined around each other's waists, while those of them nearest to the chair upon which their mother sat, leaned back against her or looked up into her face. Sophy always paused a moment before she picked out her own mother from the group of charming girls. She remembered that her mother often told her that it took the photographer twenty minutes to try and pose them, spreading out his hands, and exclaiming in Italian, and entreating them to stop laughing just for one minute. And all at once she remembered the other stories her mother used to tell her too, when she was a little girl and easily pleased, the stories of beaux and bouquets, and the stories of larks and pranks that were played upon everyone; the singing, and laughing, and the playing upon the piano that had gone on all day long. And she tried to feel sorry for her mother, all alone now, sitting here in the evenings filled with bitterness of those unfulfilled and foolish dreams.

But instead of a feeling of pity, Sophy felt an impatience and irritability—why did she marry the wrong man? But then another question formed itself in her mind as an answer to the first. Is it possible to be certain before it is too late that the man you are going to marry is the right man or not? But she was tired, and tiredness confused her mind, and she put both questions aside unanswered. Times have changed, she told herself easily. Women know more about men now than they did long ago. Marriages may break asunder nowadays, but they don't rot slowly.

"I hope you haven't given up sugar?" her mother asked. "It's so foolish of girls to try and be thinner than Nature intended them to be."

"Oh, I take sugar all right," said Sophy—and then she could

not resist a slight protest. "I'm glad I take it because if I didn't I should hate to have you try and force me to do so, Mother!"

"Am I a scold?" said her mother, smiling and seeming to enjoy the idea as a joke. She passed the jug of milk. "Put in the milk yourself. You say I always give you too much."

But Sophy was staring into the jug.

"What is the matter with the milk?" she said, and she held it up to her nose.

"There's nothing the matter with it," said her mother, anxiously looking at her.

"Ugh!" said Sophy. "There's a scum on it! Ugh, it's disgusting! What is it?"

"Oh!" said her mother relaxing in relief. "I forgot to take off the scum. But it's nothing. It's quite all right. I just gave the milk a boil while you were upstairs. It wasn't sour, but it seemed on the point of turning. Boiling saves it, but it doesn't make any difference to the milk. You wouldn't notice any taste when it's in the tea. But I shouldn't have left the scum on it. That was stupid of me!" And she dragged off the scum with the back of a spoon. It crumpled up on the side of the jug like a piece of white silk. "There's nothing disgusting about it at all!" she said again, as she saw her daughter's grimace.

"Boiled milk gives a taste to the tea," said Sophy.

"Not at all," said her mother, pouring out a cup. "I often give it a boil at night."

"Is there any milk in the kitchen that has not been boiled?" said Sophy, drawing her empty cup away as her mother went to pour the tea into it. She stood up with the cup in her hand.

"I'm afraid not," said her mother, "but I assure you this is quite all right. You won't notice the difference when it's in the tea! I promise you. I wouldn't have done it otherwise. I'm too particular about things!"

Sophy sat down and held out the cup. When it was filled she raised it to her lips and sipped at it.

"I taste it distinctly!" she said, lowering the cup again.

"You couldn't, dear." Her mother sipped hers. "I don't taste it."

"Well I do," said Sophy, "and what's more, I can't drink the tea."

"But that's absurd," cried her mother. "It's just your imagination."

"It's not imagination. I think you might have left a little milk without being boiled, when you know a cup of tea means so much to me after a long journey."

"I tell you it makes absolutely no difference."

"Oh Mother! Let's not argue about it. I can't drink the tea; that's all! Oh, why did you boil it?"

The mother sat looking at her own tea, for which she too had suddenly a great distaste. Why did she boil the milk? She tried to remember. It wasn't sour. And even if it was beginning to turn it would have kept until Sophy had come down; a matter of thirty or forty minutes. Suddenly she remembered the way she had walked from room to room during those minutes, moving the kettle, stirring the fire, and filling in the time with aimless actions.

"Perhaps if you hadn't stayed so long upstairs I might not have had time to boil it!" she said, flinging out the excuse without caring what effect it had. The evening was spoiled, anyway. The whole week of work and preparation was spoiled too. Everything was spoiled.

Sophy got up from her chair. And now she looked very tired indeed. There was no mistaking the heavy black lines under her eyes and the way her mouth dropped at the sides.

"Don't harp on it, Mother," she said, and she thought to herself that this was the way her father and mother acted when she was out of earshot.

"That's right!" said her mother. "Lose your temper now! Go upstairs and bang your door! After all the trouble I took preparing for you this is all the thanks I get. It's good that I'm used to this kind of thing from your father!"

"Leave father out of it!" cried Sophy.

"Two of a feather!" cried her mother.

Sophy threw out her hands in an appeal.

"I wish you wouldn't take it this way, Mother!"

"What other way can I take it?"

"You could admit that you made a mistake in boiling the milk without finding out whether I minded or not."

"I can't see now or then why you did mind it," said her mother.

"It makes no difference whatever to the taste of the tea. At home we always gave the milk a boil after supper to keep it from getting sour overnight."

"I never heard of it being done anywhere else," said Sophy. "You did a lot of things at home that sound queer to me, if it comes to that!"

Her mother drew in a quick breath.

"I suppose you heard your father say that! He's always having a slap at my sisters!"

"He never even mentions them, if you want to know!"

"Oh, that doesn't deceive me. There are more ways of sneering than by word of mouth. And indeed, now that we're mentioning such things, let me tell you that your own attitude isn't what it might be at times. When the old cock crows, the young cock cackles."

"Well, I can't help it, can I, if I'm like him?" said Sophy.

"There's no need to copy his ignorant traits," said her mother.

Sophy stuck her fingers into her ears.

"I won't listen to talk like that," she cried. "It's unjust. I never knew him to say or do anything that wasn't just as it should be!"

"Maybe if you were at home a little more you might not have the same thing to say for very long."

Sophy sprang to the door. "I must say I'm glad I don't come home very often if this is the kind of reception I get! Good night!" And she banged the door.

Sophy went up to the bedroom where the floor had been waxed twice over, and where the stiff muslin curtains floated back and forth as the winds urged them.

As she undressed she thought of the girls in the white dresses who vexed the photographer by laughing so much he could not pose them properly and she thought of the photograph of her father that hung in the hall, showing a stiff and straight young man with a broad face and serious eyes with a stern look in them. And suddenly she ran to her case and opened it, tossing up her blouses and handerkerchiefs as she ran her hand to the bottom and took out a small photograph in a small frame. It was a photograph of another young man, also straight and also stiff, with serious eyes and a stern look, because these are the attributes which young men wish to appear to possess when they have their photographs taken. Sophy stared at the photograph, and then she ran to the mirror and stared at her own face. But she had not learned anything from looking at either face, for she sighed and got into bed.

Two or three times she leaned up on her elbow to hear if her mother had gone up to her own room, but she heard no sound, and several times she wanted to go down again and ask to be forgiven, but she knew that instead of coming to terms they would begin to argue again, so she lay still, and began instead to plan the things that she would do for her mother when she had money of her own to spend as she liked. But even this was difficult to do because her mother and herself so rarely liked the same things. The tears came into her eyes.

And then it suddenly seemed to Sophy that she had discovered a secret, a wonderful secret, that wise men had been unable to discover, and yet it was so simple and so clear that anyone could understand it. She would go through the world teaching her message. And when it was understood there would be an end to all the misery and unhappiness, all the misunderstanding and argument with which she had been familiar all her life. Everything would be changed. Everything would be different.

The footsteps that had stopped outside her door moved on, and then were silent. A door closed far away. The dream began to form again. People would all have to become alike. They would have to look alike and speak alike and feel and talk and think alike.

What a wonderful place the world would become. People would all look alike. They would all look like the girls in the photograph, with white dresses and linked arms. They would speak and think alike. They would all think like herself and her father. It was so simple. It was so clear! She was surprised that no one had thought of it before. She saw the girls untwine their arms, and lift up the hems of their long dresses and step aside to admit her as she passed into their company.

♆

HAROLD MONRO

(E N G L I S H , 1 8 7 9 – 1 9 3 2)

Milk for the Cat

When the tea is brought at five o'clock,
And all the neat curtains are drawn with care,
The little black cat with bright green eyes
Is suddenly purring there.

At first she pretends, having nothing to do,
She has come in merely to blink by the grate,
But, though tea may be late or the milk may be sour,
She is never late.

And presently her agate eyes
Take a soft large milky haze,
And her independent casual glance
Becomes a stiff hard gaze.

Then she stamps her claws or lifts her ears
Or twists her tail and begins to stir,
Till suddenly all her little body becomes
One breathing trembling purr.

The children eat and wriggle and laugh;
The two old ladies stroke their silk:
But the cat is grown small and thin with desire,
Transformed to a creeping lust for milk.

The white saucer like some full moon descends
At last from the clouds of the table above;
She sighs and dreams and thrills and glows,
Transfigured with love.

She nestles over the shining rim,
Buries her chin in the creamy sea;
Her tail hangs loose; each drowsy paw
Is doubled under each bending knee.

A long dim ecstasy holds her life;
Her world is an infinite shapeless white,
Till her tongue has curled the last holy drop,
Then she sinks back into the night,

Draws and dips her body to heap
Her sleepy nerves in the great arm-chair,
Lies defeated and buried deep
Three or four hours unconscious there.

♟

W. H. DAVIES

(1 8 7 1 – 1 9 4 0)

The Milkmaid's Song

A Milkmaid, on a Summer's day,
Was singing, as she milked away.

The heavy, sullen cows had come
Racing when her voice called them home.

A three-legged stool, a pail that glows,
To sit and sing, and milk her cows.

Her cheeks were red, her eyes were bright,
And, like that milk, her neck was white.

The birds around her tuned their throats—
In vain—to take her perfect notes.

The cow gave up the last milk-drop,
And tarried till her song should stop.

"Wilt marry me, sweet Maid?" I said.
She laughed in scorn, and tossed her head.

And she had milked the crimson flood
E'en to my heart's last drop of blood.

Y

LIAM O'FLAHERTY

(IRISH, 1896–1984)

Milking Time

Softly, softly, the milk flowed from the taut tapering teats into its own white upward-heaving froth. It flowed from the two front teats, two white columns shooting, crossing and descending with a soft swirling movement through the billowing froth. There is no soft cadence as soothing as its sound, no scent as pure as its warm smell, cow smell, milk smell, blood smell, mingling with the thousand soft smells of a summer evening.

The cow stood on the summit of a grassy knoll. Behind her there was a rock-strewn ridge, making a grey horizon against the sky. In front there was a vast expanse of falling land, falling in flat terraces to the distant sea. Close by, the land was green-bright under the rays of the setting sun, but in the distance it was covered with a white mist, as if it rolled, dust-raising, to the sea.

The cow chewed her cud, looking through half-closed luminous eyes downwards at the mist-covered land, her red flanks shivering with content, the wanton pleasure of being milked by a sweet-smelling, crooning woman, the gentle pressure of the woman's fingers against her teats, softer than a calf's gums.

And the woman milking was in an ecstasy of happiness; for it was her first time milking her husband's cow; her cow now. They

had been married on Thursday. It was now Sunday evening and they had come together to milk, as was the custom among the people.

He lay on the grass watching her milk, listening to her crooning voice and the voices of the birds; thinking.

"Isn't it wonderful how your little fingers can milk so quickly?" he said.

She turned her head and shook her towering mass of black hair proudly; smooth-combed, winding tresses of black hair gleaming in the twilight, red lips smiling as they crooned; full white throat swelling with soft words; crooning meaningless words of joy, as she looked at him.

He looked at her joyously and smiled, swallowing his breath.

"Wasn't it lovely to-day, Kitty," he murmured, "coming from Mass?"

She bowed her head, crooning dreamily.

"Everybody was looking at us, as we came out of the chapel together. We are the tallest couple in the whole parish, and I heard several people talking about us in whispers as we passed along the road between the men sitting on the stone walls. Were you shy?"

"I was. I put my shawl out over my face, so they couldn't see me. I thought I'd never get out of sight of the people."

"After all, it's a great thing," he said.

"What's a great thing, Michael?"

His freckled face became serious. He looked away into the distance over the mist-covered falling land to where the dim horizon of the sea dwindled into a pale emptiness.

"How tall he is," she thought, "and though his arms are hard like iron, he touches me gently."

"What's a great thing, Michael?" she said again.

"Well! It's hard to say what it is, but we are here now together and there's nothing else, is there?"

"How?"

"Before, on a Sunday evening I always wanted to wander off somewhere a long way and maybe get drunk, but now I don't want anything at all only just to lie here and watch you milking the cow."

She did not reply. She flushed slightly and bent her head against the cow's warm side, thinking of other Sundays when she sat among the village women on the green hill above the beach, singing songs as they knitted. Then she too longed for something shyly, awakening, nameless longings for a gentle strong voice and the gentle pressure of strong arms.

"But," she said, "men are queer," and changing her hands she drew at the two hind teats, wetting them first with froth and pulling slowly until two fresh white streams flowed downward.

The cow raised a hoof languidly and stamped, swinging her tail. Michael laughed.

"Maybe they are," he said.

"Michael!"

"What?"

"Sure you won't be going off again on Sunday evenings to get drunk after you get tired of me?"

"I'll never get tired of you, Kitty."

"Ah, yes, it's easy saying that now when we are only a few days married, but maybe . . ."

"No, Kitty, there's going to be no maybe with us. We'll have too much work to do to get tired of one another. It's only people who have nothing to do that get tired of one another."

"It will be lovely working together, Michael. I love pulling potato stalks in autumn and then picking the potatoes off the ridge, and at dinner-time we'll roast a few in the ground with a fire of stalks."

"The two of us."

"Yes."

"But we have all summer before that. There isn't much work in summer, only fishing. I'm going fishing tomorrow."

"Then you'll be away all day and I'll be so lonely with nobody in the house."

"You won't feel it until I'll come back in the evening with a whole lot of fish. It would be grand to take you with me in the boat, but people would be laughing at us."

"Won't it be grand if you get fish? I love to spill them out of a basket on a flag and see them slipping about. And they'll be my fish now. You'll catch them for me. Oh! It is grand, Michael."

They became silent as she finished milking, passing from teat to teat, drawing the dregs, the richest of the milk. It was like a ceremony, this first milking together, initiating them into the mysterious glamour of mating; and both their minds were awed at the new strange knowledge that had come to their simple natures, something that belonged to them both, making their souls conscious of their present happiness with a dim realization of the great struggle that would follow it, struggling with the earth and with the sea for food. And this dim realization tinged their happiness with a gentle sadness, without which happiness is ever coarse and vulgar.

She finished milking, Michael rose and split half the milk into a bucket for the calf.

"You take it to him," he said, "so that he'll get used to you."

The cow lowed lazily, looking at them with great eyes; she walked with heavy hoofs to the fence beyond which her calf was waiting in a little field for his milk. Putting her head over the fence, she licked his upraised snout.

They pushed aside the cow's head and lowered the bucket to the calf. He dashed at it, sank his nozzle into the froth and began to drink greedily, his red curly back trembling with eagerness.

Kitty rubbed his forehead as he drank.

Then they walked home silently hand in hand, in the twilight.

Y

BHARTRIHARI

(INDIAN, 570?–651?)

From *The Śatakatrayam*

28

In mixture with water
The substance of milk is diluted.
Water, feeling milk's burning pain,

Yields itself to the fire's attack.
Then to relieve the plight of its friend
Milk is willing even to burn.
But mixed again with water, milk is calmed.
Compare to this the friendship of good men.

—Translated by Barbara Stoler Miller

Y

VARLAM SHALAMOV

(R U S S I A N , 1 9 0 7 – 1 9 8 2)

Condensed Milk

Envy, like all our feelings, had been dulled and weakened by hunger. We lacked the strength to experience emotions, to seek easier work, to walk, to ask, to beg. . . . We envied only our acquaintances, the ones who had been lucky enough to get office work, a job in the hospital or the stables—wherever there was none of the long physical labor glorified as heroic and noble in signs above all the camp gates. In a word, we envied only Shestakov.

External circumstances alone were capable of jolting us out of apathy and distracting us from slowly approaching death. It had to be an external and not an internal force. Inside there was only an empty scorched sensation, and we were indifferent to everything, making plans no further than the next day.

Even now I wanted to go back to the barracks and lie down on the bunk, but instead I was standing at the doors of the commissary. Purchases could be made only by petty criminals and thieves who were repeated offenders. The latter were classified as "friends of the people." There was no reason for us politicals to be there, but we couldn't take our eyes off the loaves of bread that were brown as chocolate. Our heads swam from the sweet heavy aroma of fresh bread that tickled the nostrils. I stood there, not knowing when I would find the strength within myself to return back to the barracks. I was staring at the bread when Shestakov called to me.

I'd known Shestakov on the "mainland," in Butyr Prison where we were cellmates. We weren't friends, just acquaintances. Shestakov didn't work in the mine. He was an engineer-geologist, and he was taken into the prospecting group—in the office. The lucky man barely said hello to his Moscow acquaintances. We weren't offended. Everyone looked out for himself here.

"Have a smoke," Shestakov said and he handed me a scrap of newspaper, sprinkled some tobacco on it, and lit a match, a real match.

I lit up.

"I have to talk to you," Shestakov said.

"To me?"

"Yeah."

We walked behind the barracks and sat down on the lip of the old mine. My legs immediately became heavy, but Shestakov kept swinging his new regulation-issue boots that smelled slightly of fish grease. His pant legs were rolled up, revealing checkered socks. I stared at Shestakov's feet with sincere admiration, even delight. At least one person from our cell didn't wear foot rags. Under us the ground shook from dull explosions; they were preparing the ground for the night shift. Small stones fell at our feet, rustling like unobtrusive gray birds.

"Let's go farther," said Shestakov.

"Don't worry, it won't kill us. Your socks will stay in one piece."

"That's not what I'm talking about," said Shestakov and swept his index finger along the line of the horizon. "What do you think of all that?"

"It's sure to kill us," I said. It was the last thing I wanted to think of.

"Nothing doing. I'm not willing to die."

"So?"

"I have a map," Shestakov said sluggishly. "I'll make up a group of workers, take you and we'll go to Black Springs. That's fifteen kilometers from here. I'll have a pass. And we'll make a run for the sea. Agreed?"

He recited all this as indifferently as he did quickly.

"And when we get to the sea? What then? Swim?"

"Who cares. The important thing is to begin. I can't live like this any longer. 'Better to die on your feet than live on your knees.'" Shestakov pronounced the sentence with an air of pomp. "Who said that?"

It was a familiar sentence. I tried, but lacked the strength to remember who had said those words and when. All that smacked of books was forgotten. No one believed in books.

I rolled up my pants and showed the breaks in the skin from scurvy.

"You'll be all right in the woods," said Shestakov. "Berries, vitamins. I'll lead the way. I know the road. I have a map."

I closed my eyes and thought. There were three roads to the sea from here—all of them five hundred kilometers long, no less. Even Shestakov wouldn't make it, not to mention me. Could he be taking me along as food? No, of course not. But why was he lying? He knew all that as well as I did. And suddenly I was afraid of Shestakov, the only one of us who was working in the field in which he'd been trained. Who had set him up here and at what price? Everything here had to be paid for. Either with another man's blood or another man's life.

"Okay," I said, opening my eyes. "But I need to eat and get my strength up."

"Great, great. You definitely have to do that. I'll bring you some . . . canned food. We can get it. . . ."

There are a lot of canned foods in the world—meat, fish, fruit, vegetables. . . . But best of all was condensed milk. Of course, there was no sense drinking it with hot water. You had to eat it with a spoon, smear it on bread, or swallow it slowly, from the can, eat it little by little, watching how the light liquid mass grew yellow and how a small sugar star would stick to the can. . . .

"Tomorrow," I said, choking from joy. "Condensed milk."

"Fine, fine, condensed milk." And Shestakov left.

I returned to the barracks and closed my eyes. It was hard to think. For the first time I could visualize the material nature of our psyche in all its palpability. It was painful to think, but necessary.

He'd make a group for an escape and turn everyone in. That

was crystal clear. He'd pay for his office job with our blood, with my blood. They'd either kill us there, at Black Springs, or bring us in alive and give us an extra sentence—ten or fifteen years. He couldn't help but know that there was no escape. But the milk, the condensed milk. . .

I fell asleep and in my ragged hungry dreams saw Shestakov's can of condensed milk, a monstrous can with a sky-blue label. Enormous and blue as the night sky, the can had a thousand holes punched in it, and the milk seeped out and flowed in a stream as broad as the Milky Way. My hands easily reached the sky and greedily I drank the thick, sweet, starry milk.

I don't remember what I did that day nor how I worked. I waited. I waited for the sun to set in the west and for the horses to neigh, for they guessed the end of the work day better than people.

The work horn roared hoarsely, and I set out for the barracks where I found Shestakov. He pulled two cans of condensed milk from his pockets.

I punched a hole in each of the cans with the edge of an ax, and a thick white stream flowed over the lid onto my hand.

"You should punch a second hole for the air," said Shestakov.

"That's all right," I said, licking my dirty sweet fingers.

"Let's have a spoon," said Shestakov, turning to the laborers surrounding us. Licked clean, ten glistening spoons were stretched out over the table. Everyone stood and watched as I ate. No one was indelicate about it, nor was there the slightest expectation that they might be permitted to participate. None of them could even hope that I would share this milk with them. Such things were unheard of, and their interest was absolutely selfless. I also knew that it was impossible not to stare at food disappearing in another man's mouth. I sat down so as to be comfortable and drank the milk without any bread, washing it down from time to time with cold water. I finished both cans. The audience disappeared—the show was over. Shestakov watched me with sympathy.

"You know," I said, carefully licking the spoon, "I changed my mind. Go without me."

Shestakov comprehended immediately and left without saying a word to me.

It was, of course, a weak, worthless act of vengeance just like all my feelings. But what else could I do? Warn the others? I didn't know them. But they needed a warning. Shestakov managed to convince five people. They made their escape the next week; two were killed at Black Springs and the other three stood trial a month later. Shestakov's case was considered separately "because of production considerations." He was taken away, and I met him again at a different mine six months later. He wasn't given any extra sentence for the escape attempt; the authorities played the game honestly with him even though they could have acted quite differently.

He was working in the prospecting group, was shaved and well fed, and his checkered socks were in one piece. He didn't say hello to me, but there was really no reason for him to act that way. I mean, after all, two cans of condensed milk aren't such a big deal.

—*Translated by John Glad*

EMILE VERHAEREN

(BELGIAN, 1855–1916)

The Milk

In the very low and narrow basement, next
To the vent that gets daylight from the north, the jars
Let the milk cool off in white pools,
In the red plumpness of their sandstone bellies.

To see them asleep in a dark corner, one would have thought
Of enormous lilies opening on slow currents,
Or dishes protected by white coverlids
To be reserved for angels' repasts, in the shadow.

On double planks the great barrels rested.
And the platters laden with hams and pigs knuckles,
And the white sausages bursting their wax-colored casings,

And the flans, brown with sugar along their edges,
Urging the frenzies of bellies and bodies . . .
But opposite, the milk remained cool, remained virgin.

—*Translated by Andrée-Anne Desmedt*

Y

SHERWOOD ANDERSON

(AMERICAN, 1876–1941)

Milk Bottles

I lived, during that summer, in a large room on the top floor of an old house on the North Side in Chicago. It was August and the night was hot. Until after midnight I sat—the sweat trickling down my back—under a lamp, laboring to feel my way into the lives of the fanciful people who were trying also to live in the tale on which I was at work.

It was a hopeless affair.

I became involved in the efforts of the shadowy people and they in turn became involved in the fact of the hot uncomfortable room, in the fact that, although it was what the farmers of the Middle West call "good corn-growing weather" it was plain hell to be alive in Chicago. Hand in hand the shadowy people of my fanciful world and myself groped our way through a forest in which the leaves had all been burned off the trees. The hot ground burned the shoes off our feet. We were striving to make our way through the forest and into some cool beautiful city. The fact is, as you will clearly understand, I was a little off my head.

When I gave up the struggle and got to my feet the chairs in

COFFEE, TEA, OR MILK? · 379

the room danced about. They also were running aimlessly through a burning land and striving to reach some mythical city: "I'd better get out of here and go for a walk or go jump into the lake and cool myself off," I thought.

I went down out of my room and into the street. On a lower floor of the house lived two burlesque actresses who had just come in from their evening's work and who now sat in their room talking. As I reached the street something heavy whirled past my head and broke on the stone pavement. A white liquid spurted over my clothes and the voice of one of the actresses could be heard coming from the one lighted room of the house. "Oh, hell! We live such damned lives, we do, and we work in such a town! A dog is better off! And now they are going to take booze away from us too! I come home from working in that hot theatre on a hot night like this and what do I see—a half-filled bottle of spoiled milk standing on a window sill!

"I won't stand it! I got to smash everything!" she cried.

I walked eastward from my house. From the northwestern end of the city great hordes of men, women and children had come to spend the night out of doors, by the shore of the lake. It was stifling hot there too and the air was heavy with a sense of struggle. On a few hundred acres of flat land, that had formerly been a swamp, some two million people were fighting for the peace and quiet of sleep and not getting it. Out of the half darkness, beyond the little strip of park land at the water's edge, the huge empty houses of Chicago's fashionable folk made a grayish-blue blot against the sky. "Thank the gods," I thought, "there are some people who can get out of here, who can go to the mountains or the seashore or to Europe." I stumbled in the half darkness over the legs of a woman who was lying and trying to sleep on the grass. A baby lay beside her and when she sat up it began to cry. I muttered an apology and stepped aside and as I did so my foot struck a half-filled milk bottle and I knocked it over, the milk running out on the grass. "Oh, I'm sorry. Please forgive me," I cried. "Never mind," the woman answered, "the milk is sour."

He is a tall stoop-shouldered man with prematurely grayed hair and works as a copy writer in an advertising agency in Chicago—

an agency where I also have sometimes been employed—and on that night in August I met him, walking with quick eager strides along the shore of the lake and past the tired petulant people. He did not see me at first and I wondered at the evidence of life in him when everyone else seemed half dead; but a street lamp hanging over a nearby roadway threw its light down upon my face and he pounced. "Here you, come up to my place," he cried sharply. "I've got something to show you. I was on my way down to see you. That's where I was going," he lied as he hurried me along.

We went to his apartment on a street leading back from the lake and the park. German, Polish, Italian and Jewish families, equipped with soiled blankets and the ever-present half-filled bottles of milk, had come prepared to spend the night out of doors; but the American families in the crowd were giving up the struggle to find a cool spot and a little stream of them trickled along the sidewalks, going back to hot beds in the hot houses.

It was past one o'clock and my friend's apartment was disorderly as well as hot. He explained that his wife, with their two children, had gone home to visit her mother on a farm near Springfield, Illinois.

We took off our coats and sat down. My friend's thin cheeks were flushed and his eyes shone. "You know—well—you see," he began and then hesitated and laughed like an embarrassed schoolboy. "Well now," he began again, "I've long been wanting to write something real, something besides advertisements. I suppose I'm silly but that's the way I am. It's been my dream to write something stirring and big. I suppose it's the dream of a lot of advertising writers, eh? Now look here—don't you go laughing. I think I've done it."

He explained that he had written something concerning Chicago, the capital and heart, as he said, of the whole Central West. He grew angry. "People come here from the East or from farms, or from little holes of towns like I came from and they think it smart to run Chicago into the ground," he declared. "I thought I'd show 'em up," he added, jumping up and walking nervously about the room.

He handed me many sheets of paper covered with hastily scrawled words, but I protested and asked him to read it aloud. He did, standing with his face turned away from me. There was a quiver in his voice. The thing he had written concerned some mythical town I had never seen. He called it Chicago, but in the same breath spoke of great streets flaming with color, ghost-like buildings flung up into night skies and a river, running down a path of gold into the boundless West. It was the city, I told myself, I and the people of my story had been trying to find earlier on that same evening, when because of the heat I went a little off my head and could not work any more. The people of the city he had written about were a cool-headed, brave people, marching forward to some spiritual triumph, the promise of which was inherent in the physical aspects of the town.

Now I am one who, by the careful cultivation of certain traits in my character, have succeeded in building up the more brutal side of my nature, but I cannot knock women and children down in order to get aboard Chicago streetcars, nor can I tell an author to his face that I think his work is rotten.

"You're all right, Ed. You're great. You've knocked out a regular sockdolager of a masterpiece here. Why you sound as good as Henry Mencken writing about Chicago as the literary center of America, and you've lived in Chicago and he never did. The only thing I can see you've missed is a little something about the stockyards, and you can put that in later," I added and prepared to depart.

"What's this?" I asked, picking up a half dozen sheets of paper that lay on the floor by my chair. I read it eagerly. And when I had finished reading it he stammered and apologized and then, stepping across the room, jerked the sheets out of my hand and threw them out of an open window. "I wish you hadn't seen that. It's something else I wrote about Chicago," he explained. He was flustered.

"You see the night was so hot, and, down at the office, I had to write a condensed-milk advertisement, just as I was sneaking away to come home and work on this other thing, and the streetcar was so crowded and the people stank so, and when I finally got home here—the wife being gone—the place was a mess. Well, I couldn't write and I was sore. It's been my chance, you see, the wife and kids being gone and the house being quiet. I went for a walk. I think I went a little off my head. Then I came home and wrote that stuff I've just thrown out of the window."

He grew cheerful again. "Oh, well—it's all right. Writing that fool thing stirred me up and enabled me to write this other stuff, this real stuff I showed you first, about Chicago."

And so I went home and to bed, having in this odd way stumbled upon another bit of the kind of writing that is—for better or worse—really presenting the lives of the people of these towns and cities—sometimes in prose, sometimes in stirring colorful song. It was the kind of thing Mr. Sandburg or Mr. Masters * might have done after an evening's walk on a hot night in, say, West Congress Street in Chicago.

The thing I had read of Ed's centered about a half-filled bottle of spoiled milk standing dim in the moonlight on a window sill. There had been a moon earlier on that August evening, a new moon, a thin crescent golden streak in the sky. What had happened to my friend the advertising writer was something like this— I figured it all out as I lay sleepless in bed after our talk.

I am sure I do not know whether or not it is true that all advertising writers and newspapermen want to do other kinds of writing, but Ed did, all right. The August day that had preceded the hot night had been a hard one for him to get through. All day he had been wanting to be at home in his quiet apartment producing literature, rather than sitting in an office and writing advertisements. In the late afternoon, when he had thought his desk cleared for the day, the boss of the copy writers came and ordered him to write a page advertisement for the magazines on the subject of condensed milk. "We got a chance to get a new account if we can knock out some crackerjack stuff in a hurry," he said. "I'm sorry to have to put it up to you on such a rotten hot day, Ed, but we're up against it. Let's see if you've got some of the old pep in you. Get down to hardpan now and knock out something snappy and unusual before you go home."

Ed had tried. He put away the thoughts he had been having about the city beautiful—the glowing city of the plains—and got right down to business. He thought about milk, milk for little children, the Chicagoans of the future, milk that would produce a

*Chicago writers Carl Sandburg (1878–1967) and Edgar Lee Masters (1869–1950).

little cream to put in the coffee of advertising writers in the morning, sweet fresh milk to keep all his brother and sister Chicagoans robust and strong. What Ed really wanted was a long cool drink of something with a kick in it, but he tried to make himself think he wanted a drink of milk. He gave himself over to thoughts of milk, milk condensed and yellow, milk warm from the cows his father owned when he was a boy—his mind launched a little boat and he set out on a sea of milk.

Out of it all he got what is called an original advertisement. The sea of milk on which he sailed became a mountain of cans of condensed milk, and out of that fancy he got his idea. He made a crude sketch for a picture showing wide rolling green fields with white farmhouses. Cows grazed on the green hills and at one side of the picture a barefooted boy was driving a herd of Jersey cows out of the sweet fair land and down a lane into a kind of funnel at the small end of which was a tin of the condensed milk. Over the picture he put a heading: "The health and freshness of a whole countryside is condensed into one can of Whitney-Wells Condensed Milk." The head copy writer said it was a humdinger.

And then Ed went home. He wanted to begin writing about the city beautiful at once and so didn't go out to dinner, but fished about in the ice chest and found some cold meat out of which he made himself a sandwich. Also, he poured himself a glass of milk, but it was sour. "Oh, damn!" he said and poured it into the kitchen sink.

As Ed explained to me later, he sat down and tried to begin writing his real stuff at once, but he couldn't seem to get into it. The last hour in the office, the trip home in the hot smelly car, and the taste of the sour milk in his mouth had jangled his nerves. The truth is that Ed has a rather sensitive, finely balanced nature, and it had got mussed up.

He took a walk and tried to think, but his mind wouldn't stay where he wanted it to. Ed is now a man of nearly forty and on that night his mind ran back to his young manhood in the city— and stayed there. Like other boys who had become grown men in Chicago, he had come to the city from a farm at the edge of a prairie town, and like all such town and farm boys, he had come filled with vague dreams.

What things he had hungered to do and be in Chicago! What

he had done you can fancy. For one thing he had got himself married and now lived in the apartment on the North Side. To give a real picture of his life during the twelve or fifteen years that had slipped away since he was a young man would involve writing a novel, and that is not my purpose.

Anyway, there he was in his room—come home from his walk—and it was hot and quiet and he could not manage to get into his masterpiece. How still it was in the apartment with the wife and children away! His mind stayed on the subject of his youth in the city.

He remembered a night of his young manhood when he had gone out to walk, just as he did on that August evening. Then his life wasn't complicated by the fact of the wife and children, and he lived alone in his room; but something had got on his nerves then, too. On that evening long ago he grew restless in his room and went out to walk. It was summer and first he went down by the river where ships were being loaded and then to a crowded park where girls and young fellows walked about.

He grew bold and spoke to a woman who sat alone on a park bench. She let him sit beside her and, because it was dark and she was silent, he began to talk. The night had made him sentimental. "Human beings are such hard things to get at. I wish I could get close to someone," he said. "Oh, you go on! What are you doing? You ain't trying to kid someone?" asked the woman.

Ed jumped up and walked away. He went into a long street lined with dark silent buildings and then stopped and looked about. What he wanted was to believe that in the apartment buildings were people who lived intense eager lives, who had great dreams, who were capable of great adventures. "They are really only separated from me by the brick walls," was what he told himself on that night.

It was then that the milk bottle theme first got hold of him. He went into an alleyway to look at the backs of the apartment buildings and, on that evening also, there was a moon. Its light fell upon a long row of half-filled bottles standing on window sills.

Something within him went a little sick and he hurried out of the alleyway and into the street. A man and woman walked past him and stopped before the entrance to one of the buildings.

Hoping they might be lovers, he concealed himself in the entrance to another building to listen to their conversation.

The couple turned out to be a man and wife and they were quarreling. Ed heard the woman's voice saying: "You come in here. You can't put that over on me. You say you just want to take a walk, but I know you. You want to go out and blow in some money. What I'd like to know is why you don't loosen up a little for me."

That is the story of what happened to Ed, when, as a young man, he went to walk in the city in the evening, and when he had become a man of forty and went out of his house wanting to dream and to think of a city beautiful, much the same sort of thing happened again. Perhaps the writing of the condensed-milk advertisement and the taste of the sour milk he had got out of the icebox had something to do with his mood; but, anyway, milk bottles, like a refrain in a song, got into his brain. They seemed to sit and mock at him from the windows of all the buildings in all the streets, and when he turned to look at people, he met the crowds from the West and the Northwest Sides going to the park and the lake. At the head of each little group of people marched a woman who carried a milk bottle in her hand.

And so, on that August night, Ed went home angry and disturbed, and in anger wrote of his city. Like the burlesque actress in my own house he wanted to smash something, and, as milk bottles were in his mind, he wanted to smash milk bottles. "I could grasp the neck of a milk bottle. It fits the hand so neatly. I could kill a man or woman with such a thing," he thought desperately.

He wrote, you see, the five or six sheets I had read in that mood and then felt better. And after that he wrote about the ghostlike buildings flung into the sky by the hands of a brave adventurous people and about the river that runs down a path of gold, and into the boundless West.

As you have already concluded, the city he described in his masterpiece was lifeless, but the city he, in a queer way, expressed in what he wrote about the milk bottle could not be forgotten. It frightened you a little but there it was and in spite of his anger or perhaps because of it, a lovely singing quality had got into the thing. In those few scrawled pages the miracle had been worked. I was a fool not to have put the sheets into my pocket. When I went down out of his apartment that evening I did look for them in a dark alleyway, but they had become lost in a sea of rubbish that had leaked over the tops of a long row of tin ash cans that stood at the foot of a stairway leading from the back doors of the apartments above.

♥

DIANE WAKOSKI

(AMERICAN, 1937–)

Sour Milk

You can't make it
turn sweet
again.
      Once
it was an innocent color
like the flowers of wild strawberries,
and its texture was simple
would pass through a clean cheese cloth,
its taste was fresh.
And now
with nothing more guilty than the passage of time
to chide it with,
the same substance
has turned sour and lumpy.

The sour milk
makes interesting & delicious doughs,
can be carried to a further state of bacterial action
to create new foods,
can in its own right

be considered complicated and more interesting in texture
to one who studies it closely,
like a map of all the world.

But
to most of us:
it is spoiled.
Sour.
We throw it out,
down the drain—not in the back yard—
careful not to spill any
because the smell is strong.
A good cook
would be shocked
with the waste.
But we do not live in a world of good cooks.

I am the milk.
Time passes.
You cannot make it
turn sweet
again.
I sit guiltily on the refrigerator shelf
trembling with hope for a cook
who dreams of waffles,
biscuits, dumplings
and other delicious breads
fearing the modern housewife
who will lift me off the shelf and with one deft twist
of a wrist . . .
you know the rest.

You are the milk.
When it is your turn
remember,
there is nothing more than the passage of time
we can chide you with.

Ῑ

MINH-QUAN

(VIETNAMESE, 1928-)

My Milk Goes Dry

Milk has played such an important part in my life that it should be spelled in this story in capital letters: MILK. The idea sounds laughable at first: what should milk matter to a child bereaved of her mother since the earliest infancy, and fed dairy milk with a glass bottle and a rubber nipple? The dairy milk on which I was nurtured was the Bird brand very well known in this country. But— I want to make it clear—the milk with which my life has been concerned, and which I take so much to heart, is human milk produced by human mothers.

I didn't learn the distinction between the two kinds of milk at school. The days I went to school could be counted on one's fingers, since the many odd jobs a small child could do were more useful than education. As a matter of fact, I wasn't busy all the time. I had a lot of spare time during the summer holidays, when my uncle and his wife and all my cousins went on vacation. But of course the schools too were closed at that time, so I had no hope of studying. What a nonsensical situation: when I was busy, schools were open; when I wasn't. . . . But leave it. I will tell more about it later.

To be frank, I must confess that until I became a mother for the first time, I didn't really realize the importance of nursing a baby with one's own milk. Had I brought my child up naturally and simply, as other women did, I would not have to spin a long yarn about it now. But my story is rather . . . well, I'll tell it in full.

My uncle now and then set forth ideas which sounded rather eccentric, but the small child I was then could only listen to what seemed beyond my powers of thinking and imagination.

My uncle was clearly a learned man (at least I presumed so), for he talked and talked of the East and the West, of the past and the present, with no intermission and without consulting any book. He usually opened with, "According to books," to give more emphasis to his speech and to show people that he was a man of letters and not a poor fellow.

I must point out again that my unsophisticated mind was then like a wilderness, which would have to be cleared with an axe. My uncle gave the first stroke—too hard a stroke, it seemed to me. But as I was the lowest person in the household, I could hardly protest against his judgments.

How petty and humble is the position of an infant orphan in a family! I realized it better than anyone, so I tried to make myself smaller—the smaller I could be, I thought, the better. But unfortunately, nobody would take the trouble to understand what I was doing. They took it for granted that I was disobedient, unfeeling, a liar, and had all the bad qualities possibly inherent in a naughty child.

But all the insults, and even some bloody floggings, were not so painful that they produced a sense of outrage at my uncle's half-pitying, half-depressing looks and his disheartening speeches:

"This small girl is done for!" he would exclaim to his wife. "And you still beat and scold her! What's the use of it? It only makes your hands and body tired. You see, I . . ."

He looked at me and sighed, leaving his sentence unfinished, but I understood what he would have added: "There is no hope for you, and I'm sorry for you. You are an untamed child, unable to turn over a new leaf. Even a rag can be used to wipe up dust, but you are less useful than a rag! Rubbish—no more, no less. Not only that, you are a burden to me."

I understood everything. As I stood there, my eyes would be dry, but my heart was wounded, paralyzed. Youth is generally reckless, though; it wouldn't take long for my sorrow to fade away. The next time, I would feel only surprised when my uncle unexpectedly stared at me with the same discouraging look, and began again.

"If you are . . . it's owing to those three Frenchmen," he might say.*

This comment struck me as strange indeed. What had the French

*An idiomatic reference to the French in South Vietnam.

to do with a helpless creature like me? How could they be responsible? Who knows . . . (my active imagination would work on a lot of hypotheses), had they murdered my mother, burned or plundered my house? Such a riddle was too hard for a small child to solve. It only made my head ache. Since I had nobody to ask for advice, the riddle remained a heavy unanswered burden on my heart.

After some time, I noticed that whenever my uncle talked to his friends or relatives, he constantly referred to two topics: milk, and "those three Frenchmen." Milk and Frenchmen were two topics of which he never tired; I, too, found them rather attractive.

"Look," he would say, whenever he had an audience. "These three Frenchmen are liars, dishonest people, plundering, raping, seizing other people's land. They are merciless beyond imagination, and they never flinch from any cruelty. And do you know why? Why are they so barbarous?"

Here he would stop, clear his throat, and pause so as to excite the curiosity of his listeners and make them eager for the answer. If his listeners were young men who hadn't heard this speech before, they would press him eagerly to go on. Others, like Mr. Hinh, a night watchman, would only laugh and say, "We don't know, tell us." Some, like my aunt, remained unruffled as they always were.

But whatever the attitude of his audience, my uncle always knew how to treat them. His pause was never so long that the enthusiasts cooled down, nor was it so short that his subject could be treated as other than one of momentous import.

At exactly the right moment, he would continue, "No surprise at all, my friends. It's because none of those Frenchmen is allowed to suck the breast of his mother. White men are not human beings; they are all devils. If a child is fed with animal milk, he is deprived of all human sentiments when he grows up. Believe me, I'm telling the truth. Worse yet, these barbarous people are now planning to corrupt the Annamite* in the same way."

He would give a distressing sigh, as though he were pitying the whole Annamite race thus on the road to ruin.

---

*Old name for the Vietnamese.

Once, after listening thus far to his discourse, I moved closer to him, plucked up all my courage, and asked, "Honorable uncle, may I ask you something? I did not feed on animal milk, did I? I sucked . . . I drank bird milk, and bird is not an animal; bird is a . . ."

"Oh! Blockhead! Who told you that? You fed on bird milk? A bird is just the brand-label stuck on the milk tin. It would take too long to say, for example, 'This cake is of dragon make,' so instead we say 'dragon-cake.' This does not mean the cake is made of dragon meat. It's the same with milk: the bird is merely a brand name, but the milk comes from cows. Birds never have any breast at all; even if they did, it would be too small to hold a reasonable amount of milk. And it's no easy matter to catch them, anyway. Really, I must salute your stupidity. Listen carefully (now he was stressing every word): you were drinking cow's milk. Do you understand—the milk of cows!"

I started as if struck by thunder, but before I could start to cry, my uncle began to shout. "But how dare you ask a question, you saucy girl? Children don't know anything. Really, you are too bold!"

His anger made my legs shake and my head swim, but he soon completely forgot all about his unfortunate niece. He sipped some tea to wet his throat and soon, more eloquent than ever, he was citing evidence to make his argument more convincing. For instance, he said, Emperor T'sin Tche Hwang never enjoyed the suck of his mother's breast, so he remained quite unmoved even by a "river of bloodshed" or a "mountain of corpses." But Mencius and Confucius, the celebrated philosophers, had sucked the milk of their mothers until four years of age. So had Jesus Christ. As for Buddha Shakya Muni, he was reared by his aunt since he lost his mother. (My uncle did not specify what Buddha's aunt did with her own baby while she nursed her nephew, or whether she had recourse to the dread animal milk. But I dared not voice my questions. And I heartily wished I had such an aunt.)

My uncle also cited western examples: Neron of ancient Rome, and Ulysses of Greece. The Roman emperor did not suck at his mother's breast, he said, but the Greek king did. So it was no surprise that Neron did not shrink from any cruelty, while Ulysses was a wise man.

In Annam, he added, we also had two illustrious men, genius Tran Hung and sage Trang Trinh.* Both fed on their mother's milk until they could read. In short, all the wise men, outstanding philosophers, and gentlemen of this world had sucked at their mother's breasts, and all the cruel men and merciless murderers had not.

I drank in all these words from such a wise mouth, and was almost lost in thought when suddenly my uncle looked straight at me and said:

"Furthermore, man is the highest of all living creatures. According to books, among all the beings—the flying-in-air and the running-on-earth, the four-legged are the lowest. 'As silly as an ox' is a phrase we use all the time. Yet these devil French have imported a lot of cow milk to our country, and we have been foolish enough to use it in rearing our children. The extermination of the Annamite people is only a hair's-breadth away!"

I shivered from head to toe and my hair stood on end as I listened to these words. I understood every word of his speech, even the Chinese names (although I barely knew how to hold a writing-brush). I felt overwhelmed with shame as I recalled how I had boasted of myself to my friends. I told them that if I was cleverer than all of them, it was because of the bird's milk I had drunk as a baby. Birds, I proudly said, belonged to the superior feathered creatures.

"Superior feathered creatures? What does that mean?" the children had asked, staring at me with wide eyes. I had explained as clearly as possible: "Superior means high; feathered creatures means birds, and superior feathered creatures means birds flying high in the sky, dignified and noble."

Not content with stopping there, I had gone on: "Opposite to superior feathered creatures, there are inferior hairy beings such as oxen, buffaloes, pigs, horses, etc. Inferior hairy beings are animals moving basely about on four legs, almost crawling on the ground, forever enslaved to human beings." I pointed to the buffalo on which I was riding and said, "For example, we use oxen and buf-

* Legendary Vietnamese heroes. The first defeated Kublai Khan in the thirteenth century; the second was from an important sixteenth-century northern dynasty.

faloes for plowing, and horses for riding, and pigs for meat."

The whole group of boys and girls had greatly admired my elo-
quence, and I felt so delighted that I went on lecturing, like a
professor: "Superior feathered creatures are also called 'flying
feathered creatures'; flying means moving in the air and feathered
creatures means birds. As for the inferior hairy beings, they may
also be called 'running beasts'; running means moving rapidly on
the ground."

On and on I went; as my audience showed no sign of wearying,
I talked nonstop. Now and then, I threw in an "According to
books," to impress the listeners. All at once I realized that I sounded
just like my uncle, and stopped, startled. But I had proved myself
a fine speaker. Sitting up on the buffalo's back, I had felt very
light—so light that I might almost have flown up to the sky like
a bird.

But now, my speaking career seemed to be at an end. How
could I face my friends? I felt like an animal who had once had a
fine coat of fur, but who had been shorn and now showed nothing
but an ugly skin infected with boils and itching. Until now, I had
been looked down on as a beast; from now on, I was in danger
of regarding myself as even lower than that. I felt as if the earth
were sinking under my feet and the sky darkening over my head.

But my uncle did not perceive my terrible state of mind. He
went on speaking volubly of outstanding men all over the world;
he quoted other extraordinary names, told anecdotes about them,
and offered many convincing proofs to strengthen his theory. But
I took no more pleasure in hearing his words. I confusedly heard
something about Mahatma Gandhi, and that was all. To the rest,
I was deaf.

What did Ulysses matter to me? And Neron? And even Tran
Hung Dao? Did those illustrious men pay the slightest attention
to me? Did they realize that I wished only one small thing to
comfort me—to have been reared on the milk of birds—but that
my dream was not to come true?

Suddenly, in the midst of his discourse against "those three
Frenchmen," my uncle stopped and put his finger on the forehead
of one of his daughters standing near him (he had many daughters), and shouted in a severe voice: "Mind you! If ever in the

time to come, you snobbishly follow this 'civilized way' and rear your children with cow's milk, I'll beat you to death!"

From that day on, whenever he looked at me with his deep, sharp eyes, I understood what he was thinking: "Oh, saucy girl, I take pity on you. If you become rough, unfeeling, incorrigible, it is because of your mother's early death. You had to feed at the expense of a foul, inferior, four-legged being. Now the beastly nature has infiltrated to the very core of your backbone and dominated your nature. I feel awfully sorry for you, but there is no way to save you, no way."

One day, as his talk proceeded to its vehement climax, he suddenly pointed his finger at one of his daughters and sternly shouted, "Take care of you . . ." At this, my aunt smiled sourly and cut him short: "Don't worry, your teaching has had good effects. Your niece says she will suckle her children with her own milk."

My uncle turned to her, his face crimson with anger. I really feared for my aunt, for I was sure he would scold her for her ironical remark; at the same time, I hoped he would show me some affection and praise me for my response to his preaching. But alas! he expressed his anger another way.

"You're quite a fool, aren't you?" he said to his wife. "I never mention this saucy girl; it's quite useless to teach her. I give lessons and advice to *your* children only, do you understand? But this evil girl will probably . . ."

I understood him, even though he did not finish his sentence. My face hardened and became as rough as stone. My uncle saw this, and was pleased to see his farsighted prophecy coming true so quickly. "You see?" his look of triumph seemed to say. "Look at her face right now. Nobody else could be worse than she is."

That night I wept, and I could hardly sleep. My tears, held back for the whole day, now flowed like rain. I cried until I felt relief from all my pain, and then I pondered my fate.

I regarded myself as a human being like other people; I knew I had not been transmuted into an animal. My circumstances, while unfortunate, were extraordinary, I thought: because my mother had died so early, I could hardly have done otherwise than drink bird's milk—or rather, Bird brand cow's milk. I could not deny my uncle's assertions (whatever he said had to be true), but I decided that he must have overlooked the special nature of my case. Only those who lived with their mothers but still did not suck their mothers' milk, surely, were turned into beasts by cow's milk.

Kneeling in the dark with my hands joined together, I prayed earnestly to God, Buddha, Christ, the Spirit of the Mountains, the Spirit of Rivers, and even Providence to have mercy on me and help me to safeguard my human nature. Afterwards, feeling much better, I fell into a sound sleep full of sweet dreams. I came across Ulysses and Gandhi, looking kind and noble and very much like the pictures of Sir Happiness and Sir Fortune on my uncle's Chinese calendar. One after the other, they caressed my head and said, "Don't worry, I will help you. . . ."

Suddenly a shouting voice rang in my ears and chased away my sweet dream: "You! Still lying stiff like that? Look, the sun has risen nine masts up the sky!"

I crept out of my sleeping mat, but I wasn't too frightened; "nine masts up the sky" just meant that the fifth watch was over and it was close to daybreak. Quickly I washed my face and led the buffalo to the rice field. I had not felt so calm for a long time.

Since that time, I have always prayed to Ulysses and Gandhi, as well as to the other gods, when I ask for protection.

After that day, too, I kept my ideas to myself. I had made the mistake of telling my cousins that I wanted to breast-feed my children; my words had found their way to my aunt and—worst of all—to my uncle. From now on, I would tell them nothing. I also stopped bragging about my knowledge of "superior feathered creatures" and "inferior hairy beings" to the boys and girls who drove buffalo to the fields. Soon I had almost completely stopped talking to them, too.

Instead of talking to others, I began to examine myself over and over to make sure that my actions and ideas were still human. I often caught myself looking into my own being as if it were a stranger, to search and scrutinize. I watched over myself as vigilantly as if I were an enemy. If I caught myself wishing or wanting something, I immediately checked the thought and said to myself, "Beware; the cow milk may still be lingering in your body and poisoning you."

One day my aunt told me to take some pigs to the market to

be sold. It promised to be an unpleasant job, since I should have to sit close to those grunting pigs, each one shut up in a separate bamboo crate, all the way to the market. But I was pleased to be going by myself, enjoying a little freedom. Besides, my aunt was trusting me to take the pigs safely to market and bring back the money; this must mean I was getting to be a responsible grown-up person.

As we waited on the platform for the train to leave, my aunt gave me a bright silver ten-piaster coin and said, "This is for you to buy some food on the train whenever you feel hungry. But don't be in too much of a hurry. Wait until the train stops, then look around and choose the nearest food hawker. And always get the food before paying. If you don't, the train may pull out and then you'll be without both food and money. Also, mind you don't . . ."

I stopped listening. I could only absorb so much advice at once, and wished fervently that the train would leave at once, so I could be by myself. "Mind this, mind that, what else yet?" I was muttering in a low voice, when suddenly the train rumbled and began to move. But my aunt still shouted advice from the platform: "Mind you, don't stand on the wagon steps when the train is about to start, or you'll fall. . . ."

I was rather moved by her worry, especially when she called me "my dear," as she never had before. I might have been carried away by emotion, had I not caught her next words on a last breeze: "If something happened to you, who would take care of the pigs?"

I was tempted to cry, but the giddiness of a freedom such as I had never enjoyed, and the scenery on both sides of the train, diverted me from my unhappiness. Only when the train came to periodic stops at small stations did I realize that it was quite unpleasant to ride in close quarters with pigs in a cattle car. The rest of the time, I was quite contented.

At one stop, an old woman appeared on the platform. With one hand, she leaned on a stick; with the other, she begged alms from the passengers at the train windows. Muttering unintelligible words, she waved her quivering hand up and down, back and forth, in front of the passengers for a long time. Getting no response, she moved in the direction of the cattle car, where I was. She groped her way step by step, looking quite wretched and miserable. With her eyes showing but their white corneas kept stiffly open, and with her nostrils now swelling, now deflating, and vibrating all the time, she was a sight both funny and pitiful. I felt deeply sorry for her, and put my hand in my pocket for my silver coin. I was just about to hand it out the window when I heard giggles and voices from the passenger car.

"Ha! The old woman is going to ask for alms in the cattle car! She must be blind. Nobody is there but the pigs!" But I was there, half hidden inside, and I saw two smartly dressed girls about my age lean their heads out of their coach to get a better look at the beggar; they acted as though they were enjoying a circus. I stood dead still, lest they should see me. The old woman moved on down the platform, patiently waving her hand to and fro and muttering insistently.

After a long, shrieking whistle, the train started up again. Brusquely I sat down on the floor, close to the pigs, and cried. I felt deeply grieved at the scene I had just witnessed. At the same time, I was angry with myself for having missed a rare opportunity of doing good, just because I was ashamed of being seen. I was afraid I should never have another silver coin in all the rest of my life.

Later, when I recalled that incident, I was gnawed by remorse for failing to do a good deed. But at the same time, I felt proud of myself for having acted more human than the other passengers. In fact, I later realized, the giddiness of liberty I had felt in the train had made me totally forget to keep my usual close watch on my thoughts and actions. If I had felt like doing a charitable thing, it was really and truly a natural impulse—a very noble human impulse. Having thus proved my humanity again to myself, I felt very cheerful. I wanted to shout, so that everyone could hear me, "Look! I still have a human nature in spite of having drunk cow's milk."

Of course, I didn't shout, or even speak of the incident to anyone. I had learned my lesson. If I couldn't even confide in my uncle, who was my closest relative, who could I talk to?

Even though I didn't hold a grudge against my uncle for his lack of sympathy and understanding, I was far from tolerant of everything and everybody. I hated "those three Frenchmen" terribly and relentlessly. If I was forsaken, despised, and morally tortured, who else but those French could have caused my misery? They were responsible, according to my uncle, for bringing tinned cow's milk to our country.

I was lucky enough to be gifted with a strong human nature, I told myself; otherwise, the "animal substance" of cow's milk would have supplanted it and spoiled my whole life. I vowed that if I ever had children, they would all drink only my milk. I would never let them live in such a miserable situation as I did. "Oh Lord the Mighty," I prayed. "I pray Thee not to let me suffer death so early as my mother did. Pity, pity!"

And there you are, my friends and readers. Now you know the reason for my determination to feed my children with my own milk. If I did not, they might grow silly, cruel, or inhuman, according to the wise preaching of my uncle. The story sounds very simple, but it was written with the tears of a child.

Until I had my fifth child, I kept my vow to the letter. All my first four children had drunk milk from my breasts until they were able to eat rice. Even the fifth one fed smoothly for six months, when suddenly my milk began to go dry. One morning, when I got up, my breasts felt less tight than the previous day; they remained unchanged even at feeding time, and the baby seemed unsatisfied even though it sucked longer than usual.

I tried not to think about it, but my brain worked feverishly at possible answers to the riddle. Had I wronged my breasts in any way? I had not: I abstained from wine, cigarettes, spicy dishes, even tea; ordinary rice, common sugar, peas, and kidney beans were my usual foods. At 9 p.m. I shut myself up in my room and carefully barred the door, lest my husband disturb my rest. With all that trouble just for the sake of my milk, how could I be having this trouble?

"Take it easy, my dear," said a friend of mine, who was also my midwife. "Why make a fuss over a trifle? It's natural to have less milk after several confinements. I've experienced it myself."

"Take it easy!" I retorted. "That's easy enough for you to say. You don't have my worry. I don't have enough milk to feed my child!"

She frowned. "Have you given him his bottle yet?" she asked.

"His bottle? Oh no, I don't bottle-feed him!"

"But I advised you to bottle-feed him when you were confined, didn't I?"

"Well, at first I was going to, but then I decided not to. I didn't think I needed to, since I usually have a lot of milk."

"Good heavens! Well, hats off to you, my dear, but do give him an extra bottle now. There's nothing to worry about; he's old enough to do without breast milk."

All right, since there was no better solution. But it was hard to accept her suggestion. If you gave birth to a child, I thought, it was your responsibility to feed it by yourself—otherwise what was the use of women's showy breasts?

Looking for the first time at a bottle filled with strange milk intended for one of my children, I felt vexed and angry. My hands shook and could hardly hold the bottle. The baby seemed to suck on the hard nipple unwillingly. Unaccustomed to the makeshift rubber, he tried now and then to push it out with his tongue.

As I struggled with the bottle and with my wounded pride, I suddenly remembered a time several years earlier, during the resistance fighting, when condensed milk had been as scarce as gold. While all the other mothers worried about where they would get milk, my mind was untroubled. From time to time, my husband would ask me, "Darling, won't you buy some tinned milk for the children? It may be impossible to get when we need it."

"Thanks a lot for your advice, comrade," I would teasingly reply. "But your humble companion can produce enough milk for them. Please don't worry about it; just mind your grand affairs."

Usually my jokes would make him laugh, but sometimes he continued to nag at me: "You are never serious, but the situation is, and you don't care a bit."

It wasn't that I didn't care, I thought, but what could I do about the world beyond my baby and my breast? In time of war, people had better get closer, love each other, and not fight among themselves. So I thought, but I said nothing, and gave my teat to the baby.

My husband seemed to regret his outburst; he came closer and started to talk, just as I was trying to reach a glass on a nearby table. "Do you want a drink?" he asked. "I'll make it for you." I shook my head gently and took the glass from his hand. I put it under the other nipple, and within a minute the glass was almost half full of milk. I put my head back and drank it before his amazed eyes.

"You certainly have plenty of milk, haven't you?" he said admiringly. Getting no answer, he went on, "But don't you find it unsavory?"

"Do you think I'm a fish?" I retorted. "It's my own milk!"

"Is it really good? Tell me frankly."

"Why not?" I vaguely said, adding to myself, Aha, he wants some himself! "Why do you want to know? Tell me. If you want some milk I'll buy it for you. But don't pretend to worry about the baby; you know he's never lacked enough milk. Tell me the truth."

Red-faced, he denied it: "You're joking all the time! I'm not a baby."

We both laughed. Then I said, "I lied; my milk isn't good at all; it's a bit creamy and sweet, and difficult to swallow. I was just teasing you."

"Then why did you drink it?"

"Because I didn't want to waste it. In order to recycle it for the baby, I had to drink it, and if I didn't finish it in one gulp, I might not have been able to finish it."

Remembering this conversation, I also recalled a friend's words about something called the law of compensation. I wondered whether this law applied to my present situation. When we had been poor and sometimes had to eat weeds instead of rice to ease our hunger, my milk was always plentiful; but now, when I could have anything I wanted, I was denied what I wanted most—milk.

I also thought about the other women with whom I had become acquainted in recent years. None of them breast-fed their babies. Some said it would be unhealthy, or spoil their complexions; others thought their own milk was unfit for a child's constitution. Many said it would take too much time (although they did not spend their time working, but rather strolling, shopping, or otherwise entertaining themselves). One woman said, "Whenever I let the baby feed at my breast, I feel so tired!" Another held her husband responsible: "A queer fellow, my husband. He is against my old way of nursing."

All these excuses sounded like a Greek poem to me. Although

these women were my friends, I paid no attention to their ideas on breast-feeding. Instead, I took advice from women who suggested precautions for insuring a constant flow of milk.

One of my cousins told me, "Now, drink as much milk as you can. The more milk you drink, the more you produce; just as eating pork liver is beneficial to our own livers."

I asked everyone for advice: everyone prescribed differently. My neighbors from the North suggested:

"Cook some pork trotters with papaw; that kind of food will give you a lot of milk."

"Banana flowers are the best."

"Glutinous rice soup is excellent. Eat it and your milk will be flowing like a stream."

A grand lady from Hue suggested, "There is no better thing than nenuphar seeds cooked with deer's stomach."

A friend from Saigon urged me to drink beer. Guffawing, she said, "Believe me, gulp it down and you'll have as much milk as you like. I'm not joking."

When I first experienced a shortage and had recourse to tinned milk I followed all this advice indiscriminately. I ate everything my friends advised. I even asked some religious people, though they know nothing about milk; they gave me a recipe for a decoction of fried cotton seeds, which they said was very simple but would have good effects. Since I was nervous, I not only drank the decoction, but also crunched the seeds. At first I sensed a creamy good taste, but soon got a sore throat from the seeds.

I began going to doctors, and dosed myself with every French medicine having remotely to do with lactation: mammary extract, Galactogil, Galacta syrup.

I grew heavier by the day, losing little by little my good figure, but it didn't worry me. I knew many ways of losing weight quickly, and could very easily regain my figure. Besides, my figure wasn't my biggest concern. I only wanted to be able to say that all my children had been raised only on their mother's milk, and had never drunk even a drop of tinned milk.

A sympathetic friend sent me a preserved deer foot, still coated with dirt, with instructions to cook it with kidney beans. Even this I ate, loathsome as it was. More than once, as I tried all these remedies, I could not help throwing up what I struggled to swallow.

It must be said that I accepted any and all advice because I had no organized theory of child-rearing myself. I always relied on my female intuition and on my own experience; this practice, I felt, would keep my babies healthy, less susceptible to disease, and able to recover quickly from sickness. In later years, after reading many books and articles on child-rearing, I became even more firmly convinced that this method had been the right one.

If the baby had a temperature, I would sense it in my nipple as the baby sucked; I had no need of a medical thermometer. If it caught a fever, I would try to discover the cause: a draft, too-cool bath water, or a tooth coming in? If it belched, it might be sick, or it might just have sucked too hard. If it had diarrhea, it might be because we did not cover it warmly enough; but more often, it was because we ate too many vegetables.

My love of the child, and the tender attention I paid it, gave me the ability to feel beforehand what would happen to it next. Because I had lost my own mother so early, I threw myself totally into being a mother to my own children. I learned by myself what had to be done, while I brought them up.

At the beginning, after I brought my first child into the world, I took minute care of my breasts, and they were the source of my deepest joy. While the baby fed on my milk, our mutual love grew deeper and deeper. While the baby sucked at my breast, I was listening attentively with my heart, my brain, and my senses. Overwhelmed with pride and happiness, I watched my child, that innocent infant who knew nothing but the suction of that vital motherly fluid. My heart overflowed with joy; I felt myself transported and ready to fly.

I became much stronger and steadier, feeling as if I could move rivers and mountains. I felt like a soldier waiting eagerly for the battle hour because he had entire confidence in the victory. When I held my child in my arms, all my griefs receded, and my disappointments disappeared; all the sorrows which lay heavy in my heart seemed to float away. All the hard trials of the past, the present, and the future became unimportant; I was well armed to face and overcome them.

Then one day, the child who had sucked unconsciously became

aware of me. It looked around; its brown limpid eyes, sparkling like drops of water, darted here and there. Then it turned up to stare at its mother and now and then loosed the nipple from its pretty mouth. Suddenly, it smiled, an angelic and miraculous smile. Time seemed to stand still; reclining on the bed frame, I refrained from smiling, speaking, or making the smallest movement, trying to prolong the moment, afraid that if I so much as breathed deeper, time might be startled and flap its wings more quickly.

What could be similar to that exalted state of mind? The feelings of a drunkard given a lot of wine, or of a military officer decorated for distinguished services? The immeasureable sensations of a countryman who returns home after long years abroad, or the rapture of a poet who uncovers some classical verse which has been lost for centuries?

None of these emotions seems analogous to that which I felt as I suckled my child. I cannot describe the sensation in words, but those who have experienced it will understand.

Because of the happiness it gave me—and also, perhaps, because of the lessons I learned from my uncle, I regarded milk as something of great value. When I wanted to throw out the surplus, I counted every drop. If, after a bath or before I suckled the baby, a few drops were squeezed out, I was annoyed.

I took pride in the fact that I had nursed all my children by myself—from the first one, whom I nursed in an unexperienced and awkward manner, to the fourth one, when I was better off and could hire a maid. The maid could do any of the housework except care for the children; that, I reserved for myself.

But now, with my fifth child, my record was being broken. I was rapidly growing fat from eating too much for the sake of my milk. People I met on the street began to address me in a playful manner:

"My dear Thu, you are really looking well."

"Milk has brought good effects on you as well as on your child."

"Both of you are getting equally rotund; as our saying goes, the mother grows round and the baby grows square."

These witty remarks at last aroused me, especially when I heard the giggles of my sisters-in-law. One afternoon I went surreptitiously to a weighing-machine, and watched in dismay as the needle climbed up, up—to 54 kilos. Nonsense! The machine must be out of order. I went to another druggist, and then another, but alas! every machine gave the same answer. I had become fat. All the food and remedies I had taken, instead of aiding my milk secretion, had merely stagnated in my body and made me fat.

Suddenly I was boiling with rage. According to my height, my normal weight was 48 kilos. I hadn't wanted to become fat, but only to increase my milk production. Back at home, I threw out all the remedies and foods, and returned to my normal diet.

In the seventh month after my confinement, my milk was providing only one suck a day. These were the most troublesome days yet. From dawn to dusk I spun around like a top, cleaning, boiling, and washing the bottles; heating the milk; boring holes in the rubber nipples. The holes had to be of moderate size, to pass milk through in just the right quantity. If they were too large, the suction might choke the baby; if they were too small, the bottle would be refused. I would trust no one else to make milk for the baby. I was afraid the milk might be too hot or too cold, or the water not cooked enough, or the bottle not clean, or the amount of milk over or under the ration. . . . All day long I worked, sweating through my clothes.

One day, my husband remarked. "Darling, are you practicing the arts of the conjuror? What are you muttering about all day?"

In fact, I felt very anxious about the baby's constipation. Often I started and woke up in the middle of the night. At meal time, if the maid came and signalled me with a look, I immediately put down bowl and chop-sticks and left the table. Voices followed me:

"What's so urgent?"

"My dear brother, our sister-in-law is going to investigate the baby's excrements."

"Even while eating?"

My eldest daughter said, "Mummy acts like a medical doctor, doesn't she, daddy?"

"Maybe your mother is practicing the part of Cau Tien."*

When I came back to the table, all the children would be hold-

---

*According to Chinese history, a defeated king who tried to taste the fecal matter of the Emperor to show his fidelity.

ing their noses. My husband would jokingly ask me, "Well, Sir Cau Tien, how are the excrements?" I might laugh with them, or stay sober, according to the result of my investigations.

I became a serious scholar of constipation. Books and journals on pediatrics piled up on my bed. I even read foreign books, giving my husband another opportunity to make fun of me. "In which points do you think the Vietnamese books are not so good as the others?" he would ask. To this, I remained silent, as if I didn't hear him, and went on reading.

Once I shouted at my youngest daughter, "Mai! Bring me the dictionary!" The poor girl, who did not know what a dictionary was, stood in confusion before the disordered heap of books. Once more I ordered, "Hurry up!"

"Mummy, I can't make out which is which!"

"Oh, what a stupid girl!" I reproved her, quite unthinkingly, and she began to cry. "Why are you crying? Nobody is hurting you."

Only when, sobbing, she reminded me that she had not yet been sent to kindergarten did I realize my mistake. I burst into laughter, took her in my arms, and patted her on the head to soothe her.

"I am sorry," I said. "I thought . . . Now please go quickly to my bed and pick up the biggest book. That is the dictionary."

At once I resumed my reading, standing near the stove where the kettle boiled with the milk bottle inside.

Worried as I was about constipation, nothing distressed me more than feeding with a bottle this baby who never seemed to grow accustomed to it. Now and then, he rebelled against it, twisting his body right and left, and sometimes even pushing the bottle onto the floor and spilling the milk. In the face of such a reaction, I could only helplessly cross my arms and weep. So both of us, mother and baby, cried together.

The baby was still sucking at my breast once a day, but it seemed to be no more pleasureable than bottle-feeding. Often he grumbled and grunted while eating, and sometimes, when he drew in unsuccessfully, he turned abruptly aside and gave me a smarting pain in my heart.

Finally, the day I was trying to delay arrived: the baby had nothing to suck; my milk had gone completely dry.

Seeing that I was as wretched as a chicken in heavy rain, my husband said, "You're really strange! Where did you get these ideas about breast-feeding, anyway? You treat the child as if he were a newborn, hugging and embracing him. . . . Besides, is your milk still nutritive enough to warrant mourning over it?"

Words choked in my throat; I shivered angrily from head to toe. But my husband calmly went on:

"There are plenty of tins of the best quality milk in the markets, and we can certainly afford them. Listen to me, and buy them for the baby; they would be much more nourishing."

His voice rang deafeningly in my ears at first, then seemed to get further and further away until it echoed as if from a long distance. I turned away, hid my face on the baby's shoulder, and used its clothes to dry my tears. Hugging the baby to my breast, I mourned again for the "golden days" when I could feed a baby from one breast and fill a glass from the other.

At least, I told myself, I had nursed this baby for nearly eight months without recourse to cow's milk. Surely that was the major part of the job, and I had been up to it. Still, I decided that it was probably not a good idea to have any more children.

But my sixth child came into the world in spite of my intentions. This time I was not taking anything for granted, and in the second month I began mixed feeding with one bottle each day. But the reality was worse than I imagined: in the third month, my milk began to go alarmingly dry.

Frightened, I tried every possible remedy, though not so immoderately as the time before. But alas! the better care I took of my breasts, the less they responded. I felt displeased, restless, and haunted by my failure.

In order to pull myself together, I busied myself with sewing, cleaning, sweeping, washing the linen, and bathing the children. I even read books, if I had time. But God save me from childbirth books and reviews: I had read them over and over again, and learned all their instructions by heart; I was fed up with them.

I told myself, rather unconvincingly, that milk could not stream up adequately in the daytime because I was so busy, and concentrated on nighttime. I stayed awake during the afternoon siesta time so as to sleep more soundly at night. But when night came, though I was terribly tired, I would lie in bed staring nervously at the ceiling and listening to my heart throbbing. I strained with all my senses to feel the milk surge up into my breast, but could feel nothing. Perhaps at daybreak, I would tell myself, and wait apprehensively. Some nights I heard the "odo" in the sitting-room strike every quarter-hour until the sun rose.

Finally I did not sleep at all. Often during the night I would get up, tiptoe into the next room, put a blanket over one of the children, rearrange the mosquito net over another, open a window in my husband's room. Then I would go into the sitting-room, where a very sweet moonlight showed everything at rest except my anxious and agitated self. I would sit down on a chair in a corner of the room and try to calm my nerves as I waited for time to pass. But instead, my grief would grow larger and larger, like an oil stain spreading on a sheet of paper.

After long days of anxiety and sleeplessness, I was taken ill. Always before, when I became sick, I had thrown off the sickness with my own natural vitality, without recourse to doctors and medicine. I thought I could do the same this time, but it was a big mistake. I sweated constantly, and my nose and eyes ran all the time. I felt so dizzy that when I tried to sit up I fell back down on the bed, and everything in the room seemed to turn round and round. I vomited up everything I tried to eat. For the first time in my life, I had to be helped to walk.

My grievances still troubled me, but in different ways; they changed from acute pains to more distilled, deeper sorrows. As I recovered gradually from my illness, my milk went completely dry, though I tried desperately to conserve the last few drops.

Anyhow (I tried to console myself) this baby had fed on my breast for four months. But what if there were more babies? Deprived of their mother's milk, they would lose their most effective protection.

In spite of myself, my mind kept returning to my "golden days" of milk in abundance. I shut my eyes, and saw circles of all colors dancing before me—now approaching, now receding, now very small, now growing bigger and bigger like the wavy surface of a pond where the wind blows.

Suddenly my husband's words rang in my ears: "Is your milk still nutritive enough to warrant mourning over it?"

Ah! Now your milk is. . . . What? I felt my breast hanging somewhat heavy and tense. A gleam of hope: milk? Feverishly I tried to get up, but fell back again, tired. No, not milk: only my old affliction acting up.

—Translated by Le Van Hoan

EDMUND MILLER

(1 9 4 3 –)

Winter morning
Ice in the milkbottle
Still the stars.

CHAPTER EIGHT
BEWITCHING BREWS

EUGENE O'NEILL

(AMERICAN, 1888-1953)

From *The Fountain,*
Scene i

LUIS. *(passes his hand across his eyes, then stares into the fountain dreamily)* He sang of treasure—but strange to your longing. There is in some far country of the East—Cathay, Cipango, who knows— a spot that Nature has set apart from men and blessed with peace. It is a sacred grove where all things live in the old harmony they knew before man came. Beauty resides there and is articulate. Each sound is music, and every sight a vision. The trees bear golden fruit. And in the center of the grove, there is a fountain—beautiful beyond human dreams, in whose rainbows all of life is mirrored.

In that fountain's waters, young maidens play and sing and tend it everlastingly for very joy in being one with it. This is the Fountain of Youth, he said. The wise men of that far-off land have known it many ages. They make it their last pilgrimage when sick with years and weary of their lives. Here they drink, and the years drop from them like a worn-out robe. Body and mind know youth again, and these young men, who had been old, leap up and join the handmaid's dance. Then they go back to life, but with hearts purified, and the old discords trouble them no more, but they are holy and the folk revere them. *(With a sigh)* That's his tale, my friends—but he added it is hard to find that fountain. Only to the chosen does it reveal itself.

* * *

VOICE. Life is a field
Forever growing
Beauty a fountain
Forever flowing
Upward beyond the source of sunshine

Upward beyond the azure heaven,
Born of God but
Ever returning
To merge with earth that the field may live.

JUAN PONCE DE LEON . . . What are you, Fountain? That
from which all life springs and to which it must return—God! Are
all dreams of you but the one dream? (*Bowing his head miserably*)
I do not know. Come back, Youth. Tell me this secret!

JOHN MILTON

(ENGLISH, 1608–1674)

From *Comus*

Within the navil of this hideous Wood,
Immur'd in cypress shades a Sorcerer dwels
Of *Bacchus,* and of *Circe* born, great *Comus,*
Deep skill'd in all his mothers witcheries,
And here to every thirsty wanderer,
By sly enticement gives his banefull cup,
With many murmurs mixt, whose pleasing poison
The visage quite transforms of him that drinks,
And the inglorious likenes of a beast
Fixes instead, unmoulding reasons mintage
Character'd in the face; this have I learn't
Tending my flocks hard by i'th hilly crofts,
That brow this bottom glade, whence night by night
He and his monstrous rout are heard to howl
Like stabl'd wolves, or tigers at their prey,
Doing abhorred rites to *Hecate**

* Queen of witches.

In their obscured haunts of inmost bowres.
Yet have they many baits, and guilefull spells
To inveigle and invite th'unwary sense
Of them that pass unweeting by the way.

♆

DANTE GABRIEL ROSSETTI

(E N G L I S H , 1 8 2 8 – 1 8 8 2)

For *The Wine of Circe,* by EDWARD BURNE-JONES *

Dusk-haired and gold-robed o'er the golden wine
  She stoops, wherein, distilled of death and shame,
  Sink the black drops; while, lit with fragrant flame,
Round her spread board the golden sunflowers shine.
Doth Helios here with Hecatè combine
  (O Circe, thou their votaress?) to proclaim
  'For these thy guests all rapture in Love's name,
Till pitiless Night give Day the countersign?

Lords of their hour, they come. And by her knee
  Those cowering beasts, their equals heretofore,
Wait; who with them in new equality
  To-night shall echo back the sea's dull roar
  With a vain wail from passion's tide-strown shore
Where the dishevelled seaweed hates the sea.

---

* English Pre-Raphaelite artist (1833–1898).

CAN THEMBA

(SOUTH AFRICAN, 1924–1969)

Boozers Beware of Barberton!

Number 17, Marshall Street, Ferreirastown, Johannesburg, is just about the craziest address I've met. So many people who have lived there have gone mad, even as so many other people have stood in the slummy yard drinking that poisonous brew of she- been* invention—barberton. There is an obvious connection. But the startling fact is that four of these people lived and went mad in the same room.

A South African neurologist has just sent a paper to America on barberton. He has found that it does drive certain people mad.

In the Forties, they say, a coloured woman living in this yard and selling barberton thought that her neighbours might want to take away her profitable trade in shebeen liquor. So in the dead of night she decided to "put the jinx" on this room, and here and there in the yard, by planting magic. Not long after this people began to behave strangely.

That's their story of how it all started, but the facts of what happened subsequently is hardly less startling. At about this time there lived in the fatal room, Chris Tyssen, his wife, and Willem Tyssen, his brother. Chris suddenly found out that his wife was unhappy with him. They were always quarrelling and fighting, un- til one day she just deserted him to go and live in Pretoria. This affected Chris so badly that he became very ill, and sometimes would mutter delirious nonsense. Then suddenly his brother, Wil- lem, started to act crazy. This was more serious than Chris's con- dition which was described as having had "just a touch." Willem

---

* Illicit.

did mad things like collecting bits of paper, old tins, rubbish. Then he vanished. People looked for him everywhere. Not a sign of him. Not a sound about him, for over two months. Then came the rumour that an unidentified body had been found in Germiston. He had been drowned in a lake there. The police did not suspect foul play, but his friends are still uneasy. He was given a pauper's funeral.

Chris was ejected from the room, but he still stays in the yard, sleeping outside on a miserable mattress with the stars for a blanket. He wanders in and out like a lost, bewildered animal.

But just opposite this notorious room is the Fourie room. Here lived Willie "Oom Johnnie" Fourie and his wife, Maria. Unlike many of the other victims, Oom Johnnie did not get ill at all. He was completely healthy on the night of March 17, 1957, when he suddenly got it into his head that he wanted to make a great speech. He stripped himself stark naked, grabbed an axe and jumped on the table to deliver his great oration. He made a magnificent figure standing there in his innocence and the light casting an enormous, distorted shadow on the wall and roof.

The people rushed out and sent for the Flying Squad. He was removed from there and taken to the "mad cells" at Newlands Police Station where he was duly garbed in a blanket and left among the other mixed-ups. His wife, Maria, went to see him on the following Thursday. He seemed all right. But when she went to see him again, she heard that he had died.

"Those whom the gods wish to destroy, they first make mad."

The wife is so distracted by the death of Oom Johnnie, that she's a thin edge away herself from daylight clear levelheadedness. She moves about in a daze and mumbles, "Me, I don't want to talk. I don't want to talk. I don't want to talk."

But the yard in which she lives has become the talk of Malay Camp (Ferreirastown). And in that Doom room, now occupied by two brothers, it looks like violence plans to strike again. These two brothers are very fond of each other. The other day they had been having a brotherly drink together and got a little high. When they got home the elder brother just suddenly attacked his kid brother. They had a wild fight, throwing in everything they could lay their hands on, kicking, biting, fisting. They smashed two of

the large window-panes of the Doom Room. The following day
when I got there they had contracted a sulky truce, and it was
obvious there was no longer so much bad blood between them.
But the people in the yard intimated to me in hushed whispers
that they knew it was the jinx over the room striking again.

I think, however, there is a much simpler explanation of the
phenomenon. Next to Maria Fourie's room there's another room
whose role in this business is much more ominous than people
realise. This is the house where they sell barberton in a big way.
African men come from the neighbouring mines, the town, and
industrial concerns for their mugful of *mbamba* (barberton). But
barberton is a poison made in such a way as to give a quick kick.
It is made of bread, yeast and sugar. Its main characteristic is that
it is "raw" (swiftly prepared) liquor. One of its commonest effects
is against the skin which peels off and sallows. People get red lips
and purulent black pimples on the face. But it has made those who
have drunk it for a long time raging madmen, especially in fights.
Here then may lie the answer to the mystery of the yard of lunatics. That the people have fed too long and too much on a poisonous concoction. It has made them sick and driven them mad.

Y

LUDWIG HÖLTY

(GERMAN , 1 7 4 8 - 1 7 7 6)

The Cup of Forgetfulness

One cup from the stream, where flows oblivion
through Elysium's* blooming flowers,
one cup from the stream, grant me, O genius!
There, where Sappho† forgot her youth,

*The pre-Hellenic paradise for the blessed.
†Greek lyric poet (610–580 B.C.).

there where Orpheus* forgot his Euridice,
there fill the golden urn.
Then your image shall sink, mistress imperious,
in that shimmering well of sleep!
The all-conquering glance, thrilling each vibrant nerve,
and the tremor of snowy breasts,
and the beautiful music which flowed from your lips
I shall plunge in the well of sleep.

—*Translated by J. W. Thomas*

🍷

WILLIAM SHAKESPEARE

(1 5 6 4 – 1 6 1 6)

From *Romeo and Juliet,* IV, ii

FRIAR LAURENCE.   Hold, then. Go home, be merry, give con-
   sent
To marry Paris. Wednesday is to-morrow.
To-morrow night look that thou lie alone;
Let not the nurse lie with thee in thy chamber.
Take thou this vial, being then in bed,
And this [distilled] liquor drink thou off;
When presently through all thy veins shall run
A cold and drowsy humour; for no pulse
Shall keep his native progress, but surcease;
No warmth, no [breath] shall testify thou livest;
The roses in thy lips and cheeks shall fade
To [paly] ashes, they eyes' windows fall,

---

*The legendary Greek hero used his gift of music to charm the gods into granting his wish
that he might bring his dead wife back from Hades. But in their ascent he violated the one
condition that he refrain from looking back at her, and so Orpheus lost Euridice forever.

Like death when he shuts up the day of life;
Each part, depriv'd of supple government,
Shall, stiff and stark and cold, appear like death:
And in this borrowed likeness of shrunk death
Thou shalt continue two and forty hours,
And then awake as from a pleasant sleep.
Now, when the bridegroom in the morning comes
To rouse thee from thy bed, there art thou dead.
Then, as the manner of our country is,
[In] thy best robes uncovered on the bier
Thou shall be borne to that same ancient vault
Where all the kindred of the Capulets lie.
In the mean time, against thou shalt awake,
Shall Romeo by my letters know our drift,
And hither shall he come; and he and I
Will watch thy waking, and that very night
Shall Romeo bear thee hence to Mantua.
And this shall free thee from this present shame;
If no inconstant toy, nor womanish fear,
Abate thy valour in the acting it.

Y

ERSKINE CALDWELL

(A M E R I C A N , 1 9 0 3 –)

The Medicine Man

There was nobody in Rawley who believed that Effie Henderson would ever find a man to marry her, and Effie herself had just about given up hope. But that was before the traveling herb doctor came to town.

Professor Eaton was a tall gaunt-looking man with permanent, sewn-in creases in his trousers and a high celluloid collar around his neck. He may have been ten years older than Effie, or he may have been ten years younger; it was no more easy to judge his age than it was to determine by the accent of his speech from what section of the country he had originally come.

He drove into Rawley one hot dusty morning in mid-August, selling Indian Root Tonic. Indian Root Tonic was a beady, licorice-tasting cure-all in a fancy green-blown bottle. The bottle was wrapped in a black and white label, on which the most prominent feature was the photographic reproduction of a beefy man exhibiting his expanded chest and muscles and his postage-stamp wrestler's trunks. Professor Eaton declared, and challenged any man alive to deny his statement, that his Indian Root Tonic would cure any ailment known to man, and quite a few known only to women.

Effie Henderson was the first person in town to give him a dollar for a bottle, and the first to come back for the second one.

The stand that Professor Eaton had opened up was the back seat of his mud-spattered touring car. He had paid the mayor ten ragged one-dollar bills for a permit to do business in Rawley, and he had parked his automobile in the middle of the weed-grown vacant lot behind the depot. He sold his medicine over the back seat of his car, lifting the green-blown bottles from a box at his feet as fast as the customers came up and laid down their dollars.

There had been a big crowd standing around in the weed-grown lot the evening before, but there were only a few people standing around him listening to his talk when Effie came back in the morning for her second bottle. Most of the persons there then were Negroes who did not have a dollar among them, but who had been attracted to the lot by the alcoholic fumes around the mud-caked automobile and who were willing to be convinced of Indian Root Tonic's marvelous curative powers. When Effie came up, the Negroes stepped aside, and stood at a distance watching Professor Eaton get ready to make another sale.

Effie walked up to the folded-down top in front of Professor Eaton and laid down a worn dollar bill that was as limp as a piece of wet cheesecloth.

"I just had to come back this morning for another bottle," Effie said, smiling up at Professor Eaton. "The one I took last night made me feel better than I have ever felt before in all my life.

There's not another medicine in the whole country like it, and I've tried them all, I reckon."

"Pardon me, madam," Professor Eaton said. "There are hundreds of preparations on the market today, but there is only one Indian Root Tonic. You will be doing me a great favor if you will here-after refer to my aid-to-human-life by its true and trade-marked name. Indian Root Tonic is the name of the one and only cure for ailments of any nature. It is particularly good for the mature woman, madam."

"You shouldn't call me 'madam,' Professor Eaton," Effie said, lowering her head. "I'm just a young and foolish girl, and I'm not married yet, either.'"

Professor Eaton wiped the perspiration from his upper lip and looked down at Effie.

"How utterly stupid of me, my dear young lady," he said. "Any-one can see by looking at your fresh young face that you are a mere girl. Indian Root Tonic is particularly good for the young maiden."

Effie turned around to see if any of the Negroes were close enough to hear what Professor Eaton had said. She hoped that some of the women who lived on her street would walk past the corner in time to hear Professor Eaton talk like that about her.

"I never like to talk about myself, but don't you think I am too young yet to get married, Professor Eaton?"

"My dear young lady," he continued after having paused long enough to relight his dead cigar, "Indian Root Tonic is particu-larly good for the unmarried girl. It is the greatest discovery known to medical science since the beginning of mankind. I personally secured the formula for this marvelous medicine from an old In-dian chief out in our great and glorious West, and I was com-pelled to promise him on my bended knee that I would devote the remainder of my life to traveling over this great nation of ours offering Indian Root Tonic to men and women like you who would be helpless invalids without it."

He had to pause for a moment's breath. It was then that he looked down over the folded top and for the first time looked at Effie face to face. The evening before in the glare of the gasoline torch, when the lot was crowded with people pushing and shoving to get to the medicine stand before the special introductory offer was withdrawn, he had not had time to look at everyone who came up to hand him a dollar for a bottle. But now when he looked down and saw Effie, he leaned forward to stare at her.

"Oh, Professor Eaton," Effie said, "you are such a wonderful man! Just to think that you are doing such a great work in the world!"

Professor Eaton continued to stare at Effie. She was as good-looking as the next girl in town, not over thirty, and when she fixed herself up, as she had done for nearly two hours that morning before leaving home, she usually had all the drummers in town for the day staring at her and asking the storekeepers who she was.

After a while Professor Eaton climbed out of the back seat of his car and came around to the rear where she was. He relit his cold cigar, and inspected Effie more closely.

"You know, Professor Eaton, you shouldn't talk like that to me," she said, evading his eyes. "You really don't know me well enough yet to call me 'dear girl.' This is the first time we have been alone together, and—"

"Why! I didn't think that a beautiful young girl like you would seriously object to my honorable admiration," he said, looking her up and down and screwing up his mouth when she plucked at her blouse. "It's so seldom that I have the opportunity of seeing such a charming young girl that I must have lost momentarily all sense of discretion. But, now that we are fully acquainted with each other, I'm sure you won't object to my devoted admiration. Will you?"

"Oh, Professor Eaton," Effie said excitedly, "do you really and truly think I am beautiful? So many men have told me that before, I'm accustomed to hearing it frequently, but you are the first man to say it so thrillingly!"

She tried to step backward, but she was already standing against the rear of the car. Professor Eaton moved another step closer, and there was no way for her to turn. She would not have minded that if she had not been so anxious to have a moment to look down at her blouse. She knew there must be something wrong, surely something had slipped under the waist, because Professor Eaton had not raised his eyes from her bosom since he got out of the car and came down beside her. She wondered then if she should not have confined herself when she dressed that morning, putting on all the undergarments she wore to church on Sunday morning.

"My dear girl, there is not the slightest doubt in my mind concerning your beauty. In fact, I think you are the most charming young girl it has been my good fortune to encounter during my many travels over this great country of ours—from coast to coast, from the Lakes to the Gulf."

"You make me feel so young and foolish, Professor Eaton!" Effie said, smoothing her shirtwaist over her bosom. "You make me feel like—"

Professor Eaton turned abruptly and reached into the back seat for a bottle of Indian Root Tonic. He closed his teeth over the cork stopper and popped it out, and, with no further loss of time, handed it to Effie.

"Have this one on me, my dear girl," he said. "Just drink it down, and then see if it doesn't make you feel even better still."

Effie took the green-blown bottle, looking at the picture of the strong young man in wrestler's trunks.

"I drank the whole bottle I bought last night," she said. "I drank it just before going to bed, and it made me feel so good I just couldn't lie still. I had to get up and sit on the back porch and sing awhile."

"There was never a more beneficial—"

"What particular ailment is the medicine good for, Professor Eaton?"

"Indian Root Tonic is good for whatever ails you. In fact, merely as a general conditioner it is supreme in its field. And then on the other hand, there is no complaint known to medical science that it has yet failed to allevi—to help."

Effie turned up the bottle and drank down the beady, licorice-tasting fluid, all eight ounces of it. The Negroes standing around the car looked on wistfully while the alcoholic fumes from the opened bottle drifted over the lot. Effie handed the empty bottle to Professor Eaton, after taking one last look at the picture on the label.

"Oh, Professor Eaton," she said, coming closer, "it makes me feel better already. I feel just like I was going to rise off the ground and fly away somewhere."

"Perhaps you would allow me—"

"To do what, Professor Eaton? What?"

He flicked the ashes from his cigar with the tip of his little finger.

"Perhaps you would allow me to escort you to your home," he said. "Now, it's almost dinnertime, and I was just getting ready to close up my stand until the afternoon, so if you will permit me, I'll be very glad to drive you home in my automobile. Just tell me how to get there, and we'll start right away."

"You talk so romantic, Professor Eaton," Effie said, touching his arm with her hand. "You make me feel just like a foolish young girl around you."

"Then you will permit me to see you home?"

"Of course, I will."

"Step this way, please," he said, holding open the door and taking her arm firmly in his grasp.

After they had settled themselves in the front seat, Effie turned around and looked at Professor Eaton.

"I'll bet you have had just lots and lots of love affairs with young girls like me all over the country."

"On the contrary," he said, starting the motor, "this is the first time I have ever given my serious consideration to one of your sex. You see, I apply myself faithfully to the promotion, distribution, and sale of Indian Root Tonic. But this occasion, of course, draws me willingly from the cares of business. In fact, I consider your presence in my car a great honor. I have often wished that I might—"

"And am I the first young girl—the first woman you ever courted?"

"Absolutely," he said. "Absolutely."

Professor Eaton drove out of the vacant weed-grown lot and turned the car up the street toward Effie's house. She lived only two blocks away, and during the time it took them to drive that distance neither of them spoke. Effie was busy looking out to see if people were watching her ride with Professor Eaton in his automobile, and he was busily engaged in steering through the deep white sand in the street. When they got there, Effie told him to park the machine in front of the gate where they could step out and walk directly into the house.

They got out and Effie led the way through the front door and into the parlor. She raised one of the shades a few inches and dusted off the sofa.

Professor Eaton stood near the middle of the room, looking uneasily through the small opening under the shade, and listening intently for sounds elsewhere in the house.

"Just sit down here on the sofa beside me," Effie said. "I know I am perfectly safe alone with you, Professor Eaton."

Effie closed her eyes and allowed herself the pleasure of feeling scared to death of Professor Eaton. It was an even nicer feeling than the one she had had the night before when she drank the first bottle of Indian Root Tonic and got into bed.

"And this is the ancestral home?" he asked.

"Don't let's talk about anything but you—and me," Effie said. "Wouldn't you just like to talk about us?"

Professor Eaton began to feel more at ease, now that it was evident that they were alone in the house.

"Perhaps," Professor Eaton said, sitting closer to Effie and looking down once more at her blouse, "perhaps you will permit me to diagnose your complaint. You see, I am well versed in the medical science, and I can tell you how many bottles of Indian Root Tonic you should use in your particular case. Naturally, some people require a greater number of bottles than others do."

Effie glanced out the window for a second, and then she turned to Professor Eaton.

"I won't have to—"

"Oh, no," he said, "that won't be at all necessary, though you may do as you like about it. I can just—"

"Are you sure it's perfectly all right, Professor Eaton?"

"Absolutely," he said. "Absolutely."

Effie smoothed her shirtwaist with her hands and pushed her shoulders forward. Professor Eaton bent towards her, reaching for her hand.

He held her hand for a few seconds, feeling her pulse, and then dropped it to press his ear against her bosom to listen to her heartbeat. While he listened, Effie tucked up a few loose strands of hair that had fallen over her temples.

"Perhaps," he said, raising his head momentarily, "perhaps if you will merely—"

"Of course, Professor Eaton," Effie said excitedly.

He bent closer after she had fumbled nervously with the blouse and pressed his head against her breasts. Her heartbeat jarred his eardrum.

After a while Professor Eaton sat up and loosened the knot in his necktie and wiped the perspiration from his upper lip with the back of his hand. It was warm in the room, and there was no ventilation with the door closed.

"Perhaps I have already told you—"

"Oh, no! You haven't told me!" she said eagerly, holding her hands tightly clasped and looking down at herself with bated breath. "Please go ahead and tell me, Professor Eaton!"

"Perhaps," he said, fingering the open needlework in her blouse, "perhaps you would like to know that Indian Root Tonic is the only complete aid for general health on the market today. And in addition to its general curative properties, Indian Root Tonic possesses the virtues most women find themselves in need of during the middle and later stages of life. In other words, it imparts a vital force to the glands that are in most need of new vitality. I am sure that once you discover for yourself the marvelous power of rejuvenation that Indian Root Tonic possesses, you will never again be alone in the house without it. In fact, I can say without fear of successful contradiction that—"

Effie laid her blouse aside.

"Do you want me to take—"

"Oh, yes; by all means," he replied hastily. "Now, as I was saying—"

"And this, too, Professor Eaton? This, too?"

Professor Eaton reached over and pinched her lightly. Effie giggled and passed her hands over her bosom as though she were smoothing her shirtwaist.

"I don't suppose you happen to have another bottle of that tonic in your pocket, do you, Professor Eaton?"

"I'm afraid I haven't," he said, "but just outside in my car there are several cases full. If you'll let me, I'll step out and—"

"Oh, no!" Effie cried, clutching at his arms and pulling him back beside her. "Oh, Professor Eaton, don't leave me now!"

"Very well," he said, sitting down beside her once more. "And now as I was saying, Indian Root Tonic's supernatural powers of re—"

"Professor Eaton, do you want me to take off all of this—like this?"

"Absolutely," he said. "And Indian Root Tonic has never been known to fail, whereas in so many—"

"You don't want me to leave anything—"

"Of course not. Being a doctor of the medical science, in addition to my many other activities, I need absolute freedom. Now, if you feel that you cannot place yourself entirely in my hands, perhaps it would be better if I—"

"Oh, please don't go!" Effie cried, pulling him back to the sofa beside her. "You know I have complete confidence in your abilities, Professor Eaton. I know you wouldn't—"

"Wouldn't do what?" he asked, looking down at her again.

"Oh, Professor Eaton! I'm just a young girl!"

"Well," he said, "if you are ready to place yourself entirely in my hands, I can proceed with my diagnosis. Otherwise—"

"I was only teasing you, Professor Eaton!" Effie said, squeezing his hand. "Of course I trust you. You are such a strong man, and I know you wouldn't take advantage of a weak young girl like me. If you didn't take care of me, I'd more than likely run away with myself."

"Absolutely," he said. "Now, if you will continue removing the—"

"There is only this left, Professor Eaton," Effie said. "Are you sure it will be all right?"

"Absolutely."

"But I feel so—so bare, Professor Eaton."

" 'Tis only natural to feel like that," he said, comforting her. "A young girl who has never before experienced the—"

"Experienced the what?"

"Well—as I was saying—"

"You make me feel so funny, Professor Eaton. And are you sure—"

"Absolutely," he said. "Absolutely."

"I've never felt like this before. It feels like—"

"Just place yourself completely in my hands, my dear young girl, and I promise nothing will—"

Without warning the parlor door was thrown open and Effie's brother, Burke, came in. Burke was the town marshal.

"Is dinner ready, Effie?" Burke asked, standing in the doorway and trying to accustom his eyes to the near-darkness of the parlor. "It's a quarter after twelve and—"

Burke stopped in the midst of what he was saying and stared at Effie and Professor Eaton. Effie screamed and pushed Professor Eaton away from her. He got up and stood beside Effie and the sofa, looking first at Burke and then at Effie. He did not know what to do. Effie reached for the things she had thrown aside. Professor Eaton bent down and picked up something and threw it at her.

The room suddenly appeared to Professor Eaton to be as bright as day.

"Well, I'll be damned!" Burke said, coming slowly across the floor. His holster hung from his right hip, and it swung heavily as he swayed from step to step. "I'll be damned!"

Professor Eaton stood first on one foot and then on the other. He was between Effie and her brother, and he knew of no way by which he could change his position in the room. He wished to get as far away from Effie as he possibly could. Until she had dressed herself, he hoped he would not be forced to look at her.

Burke stepped forward and pushed Professor Eaton aside. He looked at Effie and at the herb doctor, but he gave no indication of what he intended doing.

Professor Eaton shifted the weight of his body to his other foot, and Burke's hand dropped to the top of the holster, his fingers feeling for the pearl handle that protruded from it.

Effie snapped a safety pin and ran between Burke and Professor Eaton. She was still not completely dressed, but she was fully covered.

"What are you going to do, Burke?" she cried.

"That all depends on what the Professor is going to do," Burke said, still fingering the pearl handle on the pistol. "What is the Professor going to do?"

"Why, Professor Eaton and I are going to be married, Burke," she said. "Aren't we, Professor Eaton?"

"I had not intended making known the announcement of our engagement and forthcoming marriage at this time," he said, "but since we are to be married very shortly, Effie's brother should by all means be the first to know of our intentions."

"Thanks for telling me, Professor," Burke said. "It had better by a damn sight be forthcoming."

Effie ran to Professor Eaton and locked her arms around his neck.

"Oh, do you really mean it, Professor Eaton? I'm so happy I don't know what to do! But why didn't you tell me sooner that you really wanted to marry me? Do you really and truly mean it, Professor Eaton?"

"Sure," Burke said, "he means it."

"I'm the happiest girl in the whole town of Rawley," Effie cried, pressing her face against Professor Eaton's celluloid collar. "It was all so unexpected! I had never dreamed of it happening to me so soon!"

Burke backed across the room, one hand still around the pearl handle that protruded from the cowhide holster. He backed across the room and reached for the telephone receiver on the wall. He rang the central office and took the receiver from the hook.

"Hello, Janie," he said into the mouthpiece. "Ring up Reverend Edwards for me, will you, right away."

Burke leaned against the wall, looking at Effie and Professor Eaton while Janie at the central office was ringing the Reverend Edwards's number.

"Just to think that I'm going to marry a traveling herb doctor!" Effie said. "Why! all the girls in town will be so envious of me they won't speak for a month!"

"Absolutely," Professor Eaton said, pulling tight the loosened knot in his tie and adjusting it in the opening of his celluloid collar. "Absolutely. Indian Root Tonic has unlimited powers. It is undoubtedly the medical and scientific marvel of the age. Indian Root Tonic has been known to produce the most astounding results in the annals of medical history."

Effie pinned up a strand of hair that had fallen over her forehead and looked proudly upon Professor Eaton.

Y

JACK LONDON

(A M E R I C A N , 1 8 7 6 – 1 9 1 6)

From *A Hyperborean Brew*

"There I set to work. In Tummasook's copper kettle I mixed three quarts of wheat flour with five of molasses, and to this I added of water twenty quarts. Then I placed the kettle near the lamp, that it might sour in the warmth and grow strong. Moosu understood, and said my wisdom passed understanding and was greater than Solomon's, who he had heard was a wise man of old time. The kerosene can I set over the lamp, and to its nose I affixed a snout, and into the snout the bone that was like a gooseneck. I sent Moosu without to pound ice, while I connected the barrel of his gun with the gooseneck, and midway on the barrel I piled the ice he had pounded. And at the far end of the gun barrel, beyond the pan of ice, I placed a small iron pot. When the brew was strong enough (and it was two days ere it could stand on its own legs), I filled the kerosene can with it, and lighted the wicks I had braided.

"Now that all was ready, I spoke to Moosu. 'Go forth,' I said, 'to the chief men of the village, and give them greeting, and bid them come into my igloo and sleep the night away with me and the gods.'

"The brew was singing merrily when they began shoving aside the skin flap and crawling in, and I was heaping cracked ice on the gun barrel. Out of the priming hole at the far end, drip, drip, drip into the iron pot fell the liquor—*hooch,* you know. But they'd never seen the like, and giggled nervously when I made harangue about its virtues. As I talked I noted the jealousy in the shaman's eye, so when I had done, I placed him side by side with Tummasook and the woman Ipsukuk. Then I gave them to drink, and their eyes watered and their stomachs warmed, till from being afraid they reached greedily for more; and when I had them well started,

I turned to the others. Tummasook made a brag about how he had once killed a polar bear, and in the vigor of his pantomime nearly slew his mother's brother. But nobody heeded. The woman Ipsukuk fell to weeping for a son lost long years agone in the ice, and the shaman made incantation and prophecy. So it went, and before morning they were all on the floor, sleeping soundly with the gods.

"The story tells itself, does it not? The news of the magic potion spread. It was too marvellous for utterance. Tongues could tell but a tithe of the miracles it performed. It eased pain, gave surcease to sorrow, brought back old memories, dead faces, and forgotten dreams. It was a fire that ate through all the blood, and, burning, burned not. It stoutened the heart, stiffened the back, and made men more than men. It revealed the future, and gave visions and prophecy. It brimmed with wisdom and unfolded secrets. There was no end of the things it could do, and soon there was a clamoring on all hands to sleep with the gods. They brought their warmest furs, their strongest dogs, their best meats; but I sold the *hooch* with discretion, and only those were favored that brought flour and molasses and sugar. And such stores poured in that I set Moosu to build a caché to hold them, for there was soon no space in the igloo. Ere three days had passed Tummasook had gone bankrupt. The shaman, who was never more than half drunk after the first night, watched me closely and hung on for the better part of the week. But before ten days were gone even the woman Ipsukuk exhausted her provisions, and went home weak and tottery."

Y

BISHOP GEORGE BERKELEY

(IRISH, 1685 – 1753)

From *Siris*

In certain parts of America, tar-water is made by putting a quart of cold water to a quart of tar, and stirring them well together in a vessel, which is left standing till the tar sinks to the bottom. A glass of clear water, being poured off for a draught, is replaced by the same quantity of fresh water, the vessel being shaken and left to stand as before. And this is repeated for every glass, so long as the tar continues to impregnate the water sufficiently, which will appear by the smell and taste. But, as this method produceth tar-water of different degrees of strength, I choose to make it in the following manner: Pour a gallon of cold water on a quart of tar, and stir and mix them thoroughly with a ladle or flat stick, for the space of three or four minutes, after which the vessel must stand eight and forty hours, that the tar may have time to subside; when the clear water is to be poured off and kept covered for use, no more being made from the same tar, which may still serve for common purposes.

The cold infusion of tar hath been used in some of our colonies as a preservative or preparative against the small-pox, which foreign practice induced me to try it in my own neighbourhood, when the small-pox raged with great violence. And the trial fully answered my expectation, all those within my knowledge who took the tar-water having either escaped that distemper or had it very favourably. In one family there was a remarkable instance of seven children, who came all very well through the small-pox, except one young child which could not be brought to drink tar-water as the rest had done.

Several were preserved from taking the small-pox by the use of this liquor; others had it in the mildest manner; and others, that they might be able to take the infection, were obliged to intermit drinking the tar-water. I have found it may be drunk with great safety and success for any length of time, and this not only before, but also during the distemper. The general rule for taking it is, about half a pint night and morning on an empty stomach, which quantity may be varied, according to the case and age of the patient, provided it be always taken on an empty stomach, and about two hours before or after a meal. For children and squeamish persons it may be made weaker, and given little and often; more cold water or less stirring makes it weaker, as less water or more stirring makes it stronger. It should not be lighter than French, nor deeper coloured than Spanish white wine.

* * *

Cordials, vulgarly so called, act immediately on the stomach and by consent of nerves on the head. But medicines of an operation too fine and light to produce a sensible effect in the *primae viae** may, nevertheless, in their passage through the capillaries, operate on the sides of those small vessels in such manner as to quicken their oscillations, and consequently the motion of their contents, producing, in issue and effect, all the benefits of a cordial much more lasting and salutary than those of distilled spirits, which by their caustic and coagulating qualities do incomparably more mischief than good. Such a cardiac medicine is tar-water. The transient fits of mirth, produced from fermented liquors and distilled spirits, are attended with proportionable depressions of spirits in their intervals. But the calm cheerfulness arising from this water of health (as it may be justly called) is permanent. In which it emulates the virtues of that famous plant Gen Seng,† so much valued in China as the only cordial that raiseth the spirits without depressing them. Tar-water is so far from hurting the nerves, as common cordials do, that it is highly useful in cramps, spasms of the viscera, and paralytic numbness.

* * *

I do verily think there is not any other medicine whatsoever so effectual to restore a crazy constitution, and cheer a dreary mind.

---

* Alimentary canal.
† Chinese plants of the genera *Aralia* and *Phanax,* cultivated for roots that are brewed as invigorating or medicinal beverages. Both the white and red variety are in wide use in Korea as herbal "teas."

Y

(INDIAN, CIRCA 1500 B.C.)

From Rig Veda, VIII

We have drunk Soma and become immortal; we have attained
the light the Gods discovered.
Now what may foeman's malice do to harm us? What, O Im-
mortal, mortal man's deception?
Absorbed into the heart, be sweet, O Indu,* as a kind father to
his son, O Soma,
As a wise Friend to friend: do thou, wise ruler, O Soma,
lengthen out our days for living.
These glorious drops that give me freedom have I drunk.
 Closely they knit my joints as straps secure a car.
Let them protect my foot from slipping on the way: yea, let the
drops I drink preserve me from disease.
Make me shine bright like fire produced by friction: give us a
clearer sight and make us better.
For in carouse I think of thee, O Soma: Shall I, as a rich man,
attain to comfort?
May we enjoy with an enlivened spirit the juice thou givest like
ancestral riches.
O Soma, King, prolong thou our existence as Surya† makes the
shining days grow longer.
King Soma, favour us and make us prosper: we are thy devotees;
of this be mindful.
Spirit and power are fresh in us, O Indu: give us not up unto
our foeman's pleasure.
For thou hast settled in each joint, O Soma, aim of men's eyes
and guardian of our bodies.

*The Sun God.
†The river, personified and deified.

When we offend against thine holy statutes, as a kind Friend,
    God, best of all, be gracious.
May I be with the Friend whose heart is tender, who, Lord of
    Bays! when quaffed will never harm me—
This Soma now deposited within me. For this, I pray for longer
    life to Indra.*
Our maladies have lost their strength and vanished: they feared,
    and passed away into the darkness.
Soma hath risen in us, exceeding mighty, and we are come where
    men prolong existence.
Fathers, that Indu which our hearts have drunken,
Immortal in himself, hath entered mortals.
So let us serve this Soma with oblation, and rest surely in his
    grace and favour.
Associate with the Fathers thou, O Soma, hast spread
thyself abroad through earth and heaven.
So with oblation let us serve thee, Indu, and so let us
become the lords of riches,
Give us your blessing, O ye Gods, preservers. Never may sleep or
    idle talk control us.
But evermore may we, as friends of Soma, speak to the assembly
    with brave sons around us.
On all sides, Soma, thou art our life-giver: aim of all eyes, light-
    finder, come within us.
Indu, of one accord with thy protections both from
behind and from before preserve us.

                            —*Translated by R. T. H. Griffith*

---

*Chief of the Vedic gods, a warrior who brandishes thunder and is strengthened by the elixir *soma*.

♟

TOM WOLFE

(A M E R I C A N, 1 9 3 1 –)

From *The Electric Kool-Aid Acid Test*

"I looked around and people's faces were distorted . . . lights were flashing everywhere . . . the screen (sheets) at the end of the room had three or four different films on it at once, and the strobe light was flashing faster than it had been . . . the band, the Grateful Dead, was playing but I couldn't hear the music . . . people were dancing . . . someone came up to me and I shut my eyes and with a machine he projected images on the back of my eyelids (I really think this happened . . . I asked and there was such a machine) . . . and nothing was in perspective, nothing had any touch of normalcy or reality . . . I was afraid, because I honestly thought that it was all in my mind, and that I had finally flipped out.

"I sought a person I trusted, stopping and asking people what was happening . . . mostly they laughed, not believing that I didn't know. I found a man I knew not very well but with whom I felt simpatico from the first time we met. I asked him what was happening, and if it was all me, and he laughed and held me very close and told me that the Kool-Aid had been 'spiked' and that I was just beginning my first LSD* experience . . . and not to be afraid, but to neither accept nor reject . . . to always keep open, not to struggle or try to make it stop. He held me for a long time and we grew closer than two people can be . . . our bones merged, our skin was one skin, there was no place where we could separate, where he stopped and I began. This closeness is impossible to describe in any but melodramatic terms . . . still, I did feel that we had merged and become one in the true sense, that there was

*D lysergic acid diethylamide—a popular experimental drug.

nothing that could separate us, and that it had meaning beyond anything that had ever been. (Note, a year and two months later . . . three months . . . I later read about 'imprint' and that it was possible that we would continue to be meaningful to each other no matter what circumstances . . . I think this is true . . . the person in question remains very special in my life, and I in his, though we have no contact and see each other infrequently . . . we share something that will last. Oh hell! There's no way to talk about that without sounding goopy.)

"I wasn't afraid anymore and started to look around. The setting for the above scene had been the smaller room which was illuminated only by black light, which turns people into beautiful color and texture. I saw about 10 people sitting directly under the black light, which was back-draped by a white (luminescent lavender, then) sheet, painting on disembodied mannequins with fluorescent paint . . . and on each other, their clothes, etc. I stood under the light and drops of paint fell on my foot and sandal, and it was exquisite. I returned to this light frequently . . . it was peaceful and beautiful beyond description. My skin had depth and texture under the light . . . a velvety purple. I remember wishing it could be that color always. (I still do.)

"There was much activity in the large room. People were dancing and the band was playing—but I couldn't hear them. I can't remember a note of the music, because the vibrations were so intense. I am music-oriented—sing, play instruments, etc.—which is why this seems unusual to me. I stood close to the band and let the vibrations engulf me. They started in my toes and every inch of me was quivering with them . . . they made a journey through my nervous system (I remember picturing myself as one of the charts we had studied in biology which shows the nerve network), traveling each tiny path, finally reaching the top of my head, where they exploded in glorious patterns of color and line . . . perhaps like a Steinberg* cartoon? . . . I remember intense colors, but always with black lines . . . not exactly patterns, but with some outlines and definitions.

"The strobe light broke midway . . . I think they blew some-

*Saul Steinberg (1914–), known for his work in *The New Yorker*.

thing in it . . . but that was a relief, because I had been drawn to it but it disturbed the part of me that was trying to hang on to reality . . . playing with time-sense was something I'd never done . . . and I found it irresistible but frightening.

"The Kool-Aid had been served at 10 or so. Almost from the first the doorway was crowded with people walking in and out, and policemen. There were, throughout the evening, at least six different groups of police . . . starting with the Compton City police, then the Highway Patrol, sheriff's deputies, L.A.P.D. and the vice/narco squad. I seem to remember them in groups of five or six, standing just inside the doorway, watching, sometimes talking to passers-by, but making no hostile gestures or threatening statements. It seems now that they must have realized that whatever was going on was more than could be coped with . . . and a jail full of 150 people on acid was infinitely undesirable . . . so they'd look, comment, go away, and others would come . . . this continued through the night."

Y

JOHN DIGBY

(1 9 3 8 –)

The One and Only Bottle

No one around in the district could understand why my next-door neighbor, Miles O'Sheen, let his farm go to seed. His whole spread was really in a sad state, and he had had a good herd of cattle together with a decent herd of dairy cows.

Obviously the man had troubles; his farm hands quit and his livestock were gradually dying. His wife, a pretty young woman but rather on the heavy side, had left him a couple of years back. Maybe this was the reason why everything was in such a sad state of neglect.

"Of course not, a man like that can get a woman any time," my

wife answered as she stacked the washed dishes away on the shelf.

"He's still pretty young and not bad-looking," I offered.

"If he'd quit the bottle he might pull himself together," she said.

"If he doesn't I wouldn't mind putting in an offer for his spread," I repeated to myself as the thought wandered through my head.

"Christ, not while he's in that sorry state," Mary said, shaking her head in disbelief and looking at me quizzically.

"Anyway, I'll take a drive over there and chat with him. Maybe he needs the company."

"Invite him to dinner. I think he could do with a solid meal inside him."

"That I will."

In the afternoon I drove over to his farm. I had the idea that he would be sitting at the back of his house under an elm tree at a small painted iron table drinking as usual. I thought to myself then he must be in a continuous state of near drunkenness.

I parked my car outside the front of his house and walked around to the back where he was lolling on his chair with his feet spread up on the table. His bottle was there and next to it a couple of glasses. I thought it strange that he had two glasses, as if he were expecting me to call on him that afternoon.

"Afternoon Miles," I said as I approached the table and sat down before he offered the chair.

"Grab yourself a drink," he smiled, and pushed the bottle toward me.

"It's a little early for me," I said.

"Nonsense," he uttered, waving the bottle around in the air. "A nip won't harm anyone."

I sat there a little uneasy, trying to smile.

"What will it be? You fancy a gin, beer, whiskey, ale? Anything you like."

"Thanks, no. Anyway I see only one bottle."

"That's all we need," he said.

He held the bottle in his hands cupping them around it and smiling.

"Try a Tom Collins, a rum or a double whiskey. I really mean it, anything you like."

Obviously the man was drunk. I was in half a mind to get up

and depart. I was then thinking that he was a hopeless case. It would be a waste of time talking to him.

"Listen, I really mean it . . . anything you like from this little ole' bottle," he said, holding it as if it were the breast of a woman.

I think the man was challenging me. I weakened and accepted the drink.

"I'll try a double rum," I said, offering him my glass as if it were a peace offering.

"A double rum? Coming up."

He poured a couple of inches into my glass. His hand was none too steady.

"Drink it, the best rum you'll ever taste."

I held the glass up to my nose. The almost-colorless liquid had no smell whatsoever. I was considering that it might be a home-made brew of his own concoction. I sipped it. Christ, it was a rum! He threw back his head and laughed.

"Well, what do you think of that for a drink?"

Frankly, I was puzzled. I sniffed the colorless liquid again. There was no aroma.

"You fancy a Bloody Mary, a Whiskey Sour, a Sidecar?"

I looked at him with mild astonishment.

"Empty your glass, go on man, empty your glass," he said.

I threw the remains of my drink on the grass.

"One Bloody Mary coming up."

I sipped it. And it was a Bloody Mary. I held the glass up in the air, endeavoring to solve the mystery or to work out the trick, for the drink he had poured was as colorless as the one before. He must have read my mind.

"There's no trick," he offered, "anything you like as endless and bottomless as the sea. Just say the drink and then it appears."

I couldn't or wouldn't believe it. There must be a trick to it somewhere. I asked for an exotic drink. Surely this would stump him.

"Give me a Sazerac," I said, stretching out my empty glass. Silently he poured the liquid. I tasted it. Yes, it was Sazerac all right.

"I just can't believe it."

"Nor could I at first," he said sadly.

"But . . . ?"

"How? The story is simple enough. A couple of years ago a peddlar came knocking at the farm door. It was a rainy night and none too warm. She asked for a few scraps, something to eat. I fed her and gave her an old overcoat that I had no use for. At first I never thought anything of it, but she wouldn't accept a drink from me. She kept taking nips from a bottle she was holding. As she drank the merrier she became. I was fascinated. I thought to myself, it must be pretty good stuff the old girl is knocking down. She offered me a drink and like a fool I purchased the bottle. Now I can't get rid of the damn thing."

"Just get rid of it," I said.

"It's not that easy," he answered. "What man would take it?"

"Give it to me and I'll destroy it."

"No," he answered firmly. "We all believe we can control our-selves . . . but you too will slide downhill and end in the same worry-state that I have. Believe me, there's a damn curse on this bottle. Once you have it there's only one way, and that's down."

I reached over and took the bottle in my hand. He was becoming nervous.

"I would never have believed it," I said, looking hard at the bottle still trying to work out the mystery.

"Listen," he said, "forget the story and have another drink."

"For God's sake, man, destroy it before it destroys you," I said.

He shook his head, poured himself another drink, and muttered something which I couldn't catch.

"No. It's not that easy."

"Just pick it up and smash it."

"Are you crazy? No . . . no . . . no, . . ." he answered hur-riedly. "She said that if the bottle was broken it would rain forever . . . until the end of time. We would all drown."

"Miles, you can't possibly believe that rubbish, can you?"

"You don't understand. I do believe it."

I reached over quickly, picked up the bottle and with one swift movement brought it down on the edge of the wrought-iron ta-ble. It was shattered. The liquid seeped quickly into the uncut grass and disappeared.

He looked at me with disbelief, a sort of horror crawled across his face. We were both silent. We waited.

"There, where's your rain?"

I hadn't seen a man cry for years. He held his head in his hands and sobbed.

"You fool, you fool, . . ." he repeated between the sobs. "You've . . . you've destroyed us, destroyed everything."

"Miles, how can you believe such trash? You've got to take hold of yourself."

He wasn't listening to me. It was pointless to continue our conversation. I got up to leave. I couldn't very well bring him over to dinner in this state. I touched his shoulder, he brushed my hand away.

"Tomorrow you'll thank me," I said as I left him to drive home. "I'll call tomorrow," I reiterated, offering reassurances as I walked toward the car.

As I was about to drive away he shouted between his tears, "There won't be a tomorrow you damn fool, you interfering bastard."

I ignored his remarks and started to pull out from his farm.

"Tomorrow," I shouted, "the same time. Good-bye Miles."

I couldn't have gone more than a couple of miles when the sky grew dark and a light rain began to fall. Almost without warning the rain started falling in torrents. It became so heavy that I had to pull the car over to the side of the road as I had little or no visibility. The curtain of rain grew darker. The rain hammered on the car roof as loudly as a thousand African drums.

It was raining and raining heavily.

♈

PHILIP FRENEAU

(1 7 5 2 – 1 8 3 2)

The Parting Glass

The man that joins in life's career
And hopes to find some comfort here;
To rise above this earthly mass,
The only way's to drink his GLASS.

But, still, on this uncertain stage,
Where hopes and fears the soul engage;
And while, amid the joyous band,
Unheeded flows the measured sand,
Forget not as the moments pass,
That TIME *shall bring the parting glass!*

In spite of all the mirth I've heard,
This is the glass I always feared;
The glass that would the rest destroy,
The farewell cup, the close of joy!

With YOU whom Reason taught to *think*,
I could, for ages, sit and drink:
But with the fool, the sot, the ass,
I haste to take the parting glass.

The luckless wight, that still delays
His draught of joys to future days,
Delays too long—for then, alas!
Old age steps up, and—breaks the glass!

The nymph, who boasts no borrowed charms,
Whose sprightly wit my fancy warms;

What tho' she tends this country inn,
And mixes wine, and deals out *gin?*
With such a kind, obliging lass
I sigh, to take the parting glass.

With him, who always talks of gain,
(Dull Momus,* of the plodding train)—
The wretch, who thrives by others' woes,
And carries grief where'er he goes:—
With people of this knavish class
The first is still my parting glass.

With those that drink before they dine—
With him that apes the grunting swine,
Who fills his page with low abuse,
And strives to act the gabbling goose
Turned out by fate to feed on grass—
Boy, give me quick, the parting glass.

The man, whose friendship is sincere,
Who knows no guilt, and feels no fear:—
It would require a heart of brass
With him to take the parting glass!

With him, who quaffs his pot of ale;
Who holds to all an even scale;
Who hates a knave, in each disguise,
And fears him not—whate'er his size—
With him, well pleased my days to pass,
May heaven forbid the PARTING GLASS!

---

*God of ridicule, the son of night.

A GLOSSARY OF DRINKS

Absinthe   A distilled liquor flavored with licorice, hyssop, fennel, angelica, aniseed, and staranise, along with toxic wormwood (*Artemisia*). When mixed with water, the yellow-green liquor turns opalescent-white. Henry-Louis Pernod, who first produced absinthe in 1797, gave his own name to a substitute produced without wormwood. Absinthe is now illegal in France, where Pernod and Ricard, another substitute produced by M. Pernod, are commonly drunk.

Ale   A drink made from barley that is spread on the granary floor, then wetted and germinated. When it is kiln-dried it becomes malt, which is then fermented in a fresh brewing process without hops. Water and yeast are added; fermentation transforms the malt sugar into carbonic acid gas, which escapes from the liquid, leaving ale.

Alexander   A cocktail made from 1 ounce of gin, ¾ ounce creme de cacao, and ¾ ounce heavy cream. The "Brandy Alexander" substitutes brandy for the gin.

"Allegant"   English corruption of Alicante, a province in Spain known for a sweet red dessert wine of the same name.

Amontillado   A *montilla*-type sherry made from Palomino grapes. It is dry, nutty, and quite complex.

Angostura bitters   Concentrated bitter flavoring from Trinidad, used in mixed drinks. The generic term "bitters" refers to a variety of similar tinctures.

Apéritif   An alcoholic drink taken as an appetizer to a meal.

Aquavit   A colorless spirit distilled from grain or potatoes and flavored with caraway seed.

Aqua vitae   Literally "the water of life," used by the French to refer to brandy as a "medicinal," and to strong liquor generally.

Arrack   The name of several different spirits. One sort refers to a drink made from fermented cocoa-palm sap or molasses that was brought to Europe by Dutch and German traders. Batavian arrack was the name given to the Dutch import from its colony, Batavia. Moguls and Tartars use the same name to refer to a drink distilled from fermented mare's milk. In Asia Minor, arrack is a brandy flavored with anise and coriander. The colorless liquid turns milky-white when diluted with water.

Bacardi A cocktail made from 1 jigger of rum (Bacardi), ½ teaspoon granulated sugar, the juice of ½ small lime, and dash of grenadine. These are shaken with crushed ice and strained.

Barley water A non-alcoholic grain beverage made by boiling the grain in water and cooling it to room temperature. Popular in Victorian England, the drink commonly is served with meals in present-day Korea.

Beaujolais A region of France that gives its name to fruity wines derived principally from the Gamay grape. The classic Beaujolais is a red wine, though a white version exists.

Beer A drink fermented from malted and hopped barley and then filtered. Its alcoholic strength depends on the quality of malt and water as well as the individual manner of brewing. The hops that distinguish ale from beer, and give European beers a distinctly bitter flavor, were introduced from Europe to England in the sixteenth century.

Benedictine A liqueur made originally by monks of the Benedictine order at Frécamp, Normandy, according to a secret formula that included thirty-five aromatic seeds, spices, and herbs. These are distilled first in pure spirits, then redistilled with the addition of honey, cognac, wine, and caramel. Modern production passed to a commercial firm as the result of an inheritance from a former treasurer of the abbey. The hallmark "DOM" on the label is an abbreviation for *Deo Optimo Maximo* (To God, Most Good, Most Great).

Bitter An English term for a strong, dark draft beer. The term as a plural, "bitters," refers to any number of concentrated fruit and herbal flavorings, originally medicinal, that are used to enhance drinks.

Bloody Mary A long drink made by pouring 1 jigger of vodka and the juice of ½ lemon over ice, then filling the glass with tomato juice seasoned with Worcestershire sauce, horseradish, salt, pepper, and Tabasco to taste.

Bock The French term for a half pint of light beer served in a tankard. "Bock" also refers to a dark, sweet German beer brewed in winter for consumption in the spring.

Bordeaux The most important wine region of France, located in the Gironde Département. It was part of a dowry that Eleanor of Aquitaine brought with her marriage to Henry II in 1152, and as a result wines from this region were popular in England very early. Bordeaux wines are made from Merlot, Malbec, Semillon, and Sauvignon (generally recognized as the best) grapes. The generic name implies a less distinguished wine than those of the same region that carry the name of their commune, such as Margaux, Saint-Estèphe, Saint-Julien, Saint-Émilion.

Bourbon A grain spirit distilled from at least 51 percent maize.

Brandy A spirit distilled from wine.

Bronx A cocktail named after the Bronx Zoo by its inventor at the old Waldorf-Astoria, Johnnie Salon. The original contained ⅔ gin, ⅓ orange juice, to which dashes of dry and sweet vermouth were added. It was altered to ½ gin, ¼ dry, and ¼ sweet vermouth with a twist of orange peel.

Burgundy A major French wine-growing province that produces red and white wines that use the generic name of the region. The three provincial départements are known for different types of wines: Yonne for Chablis; Côte d'Or for red and white "burgundies"; Saône-et-Loire for Mâçon and Beaujolais.

Cabernet The family name of several types of black grapes.

Canary A sweet white dessert wine from the Canary Islands.

Chablis A dry white wine named after its village of origin in the Yonne Département. It is made from Chardonnay grapes. The successful growth of these grapes in the United States has led to the wide popularity of domestic Chablis as a table wine.

Chamomile A plant of the *Anthemis* genus, especially *A. nobilis,* with bitter, aromatic flowers and scented foliage brewed as an antispasmodic infusion or calming "tea."

Champagne The most famous of all sparkling wines, named for its region of production in northeast France, about ninety miles from Paris. The region's extremely chalky soil contributes to the special flavor of this wine, which is made from a blend of Chardonnay grapes (also used in white burgundy) and Pinot Noir grapes (used in red burgundy). Champagne is one of the few distinguished wines that is a conscious blend. It is made by a double fermentation process, during which the bubbles of carbon dioxide released in fermentation are retained in the bottle, causing a natural sparkle. The discovery of this process is sometimes attributed to the monastic cellar master, Dom Perignon (who, in the late seventeenth or early eighteenth century also developed the prototype of cork stoppers that fasten to bottle necks). Another theory argues that the English discovered sparkling wine accidentally when importing the regional wine during its second fermentation process.

Chartreuse A liqueur made originally by Carthusian monks of La Grande

Chartreuse near Grenoble, who began to produce the liqueur in 1607. There are two varieties: yellow and green, both containing over a hundred plant ingredients. Production is now commercial.

Châteauneuf-du-Pape A small town on the left bank of the Rhône near Avignon, known for its red wine.

Chian wines Wines from the island of Chios in Greece, highly praised in antiquity.

Chinon A red wine of the Loire Valley made from Cabernet Franc grapes. Its town of origin is also the birthplace of Rabelais.

Cider A beverage made from pressed apple juice that may or may not be fermented. There are both still and sparkling varieties.

Claret From the twelfth century, the English name for the red wines of Bordeaux.

Cocoa A chocolate drink made from cocoa beans. In 1828 Conrad J. van Houten developed a process of making chocolate powder by pressing vegetable fat from the liquor paste prepared from cocoa beans. The "Dutch Process" includes an alkali treatment that reduces acidity and enhances flavor.

Coffee A brew made from roasted beans of the plant genus *Coffea*. Twenty-five species are cultivated for beans, most varieties of *C. arabica*, originally from Arabia and now grown in South America; and *C. robusta*, originally from East Africa, now widely grown across the continent. Both varieties are cultivated in Asia, along with newer subspecies. The name is thought to originate either from the Arabic *quhwah*, or the Ethiopian *kaffa*.

Cointreau A colorless, orange-flavored liqueur.

Curaçao An orange-flavored liqueur that originated in Holland, distilled from bitter oranges found on the Dutch island-colony off Venezuela.

Daiquiri A cocktail made by adding the juice of ½ lime and a teaspoon of granulated sugar to a jigger of rum. Since the invention of prepared daiquiri mixes, flavored versions, such as the strawberry daiquiri, are common variations.

Demijohn A narrow-necked 1–10-gallon bottle, covered in wickerwork and used to store homemade whiskey.

Dheno A Bengali term for a liquor distilled from unmilled rough rice (paddy).

Eggnog A rich drink made from the yolks and whites of 8 eggs; these are beaten separately, the whites with ½ pound sugar until stiff. The two mixtures are then folded together and laced with a fifth of rum or a combination of rum and whiskey. Then a pint of whipped cream is floated on top and garnished with nutmeg. Generally regarded as a Christmas-season drink, the commercial preparation is a poor substitute!

Falernian wine An Italian wine from Falernum in the district of Campania, frequently celebrated by the ancient Roman poets.

Frumenty An English country drink made of sweetened grain and milk, laced with spirits.

Gin A distilled spirit flavored with juniper berries, *genièvre,* from which its name is derived. Dutch gins are still called "Geneva."

Gin Fizz A drink that begins with 4 ounces of gin poured over shaved ice, to which are added the juice of ½ lemon, powdered sugar to taste, and a pinch of baking soda for effervescence.

Ginger beer A sweet, carbonated beverage flavored with ginger, capiscum, or both.

Graves Both red and white wines from a district in the Gironde Département north of Bordeaux.

Grog An alcoholic liquor, usually rum, diluted with water.

Harvey Wallbanger A long drink made from 1 jigger of vodka and ½ ounce Galliano liqueur stirred on ice with 6 ounces of orange juice.

Haut-Brion A red wine of the Graves region, the Médoc district.

Highball A long iced drink made with 1 jigger of whiskey with club soda, ginger ale or water, and a twist of lemon.

Hock An eighteenth-century synonym for the older "rhenish." Both words are English terms for Rhine wines.

Hooch Derived from "Hoochinoo," a Tlingit people of Alaska who distilled liquor illegally, the word refers generally to illicit liquor.

Irish whiskey A spirit distilled in a pot still, principally from malted barley. The malt is dried in a solid floor kiln so that smoke from the fuel does not make direct contact with the grain. This differs from "smoke-cured" Scotch whiskey.

Jigger A measure of 1 ½ ounces used in mixing drinks.

King Alfonse A sweet liqueur drink made by filling a cordial glass ¾ full with dark crème de cacao and floating heavy cream on top.

Kirsch A spirit distilled in Germany, Alsace, and Switzerland from wild mountain black cherries.

Kümmel A spirit distilled from grain or potatoes and flavored with caraway and cumin seeds. Produced in Holland since 1575, it is also made in Riga and Berlin.

Lacryma Christi A golden, sweet, and delicate wine made from grapes grown on the lower slopes of Mount Vesuvius.

Liqueur A flavored and sweetened spirit which may be made from brandy, gin, rum, whiskey, neutral alcohol, as well as potatoes or sawdust. Distinct flavors and aromas are derived from flowers, roots, seeds, leaves, fruit, or a mixture of several.

Long Island Iced Tea A long iced drink that is a true endurance test. Pour over ice ¾ ounce each of vodka, gin, tequila, Triple Sec, and white rum. Add 2 ounces "Sour Mix" and 2 ounces of Coca-Cola.

Madeira Red wines fortified with brandy, produced on the island of Madeira. Like sherries, they range from dry (sercial) and rich (bual) to sweet (verdelho and malmsey).

Malaga A type of sweet Spanish wine named after the shipping port in the province of Eastern Andalucia. The best are made from Muscat grapes.

Malmsey A sweet dessert wine made from the Malvasia grape. Generally thought of as a Madeira, the particular species of grape from which it is made originated on Crete and spread to cultivation on Majorca, Cyprus, and the Canary Islands, as well as Madeira. The wine was imported to England by the fifteenth century and became very popular.

Manhattan A cocktail made from 1 jigger of rye and ½ jigger of sweet vermouth to which are added a dash of Angostura bitters and a maraschino cherry.

Manzanilla A pale dry sherry from the Sanlucar vineyards near Jerez.

Margarita A cocktail made from ½ ounce tequila, ½ ounce Triple Sec (blended orange liqueur), and the juice of ½ lime. These are shaken together over ice and poured into a glass rimmed with coarse salt.

Marsala A versatile dessert wine that is also used in cookery, made from grapes grown between Palermo and Messina in Sicily. The generic name refers to the town near its growing region.

Martini A cocktail that was first made by mixing gin and dry vermouth in equal proportions. Progressively the proportion of gin to vermouth increased, and aficionados of the "dry" Martini insist that only a suggestion of vermouth is necessary. While the cocktail was once flavored with orange bitters, a twist of lemon peel is common now, as is a green olive.

Médoc An area fifty miles long but only five miles wide on the left bank of the Gironde known for its sixty classed growths of red wine, among which are Lafite, La Tour, Haut-Brion, and Mouton-Rothschild.

Metheglin The anglicized Welsh name for a honey drink.

Mickey Finn Whiskey that has been doctored with a purgative or "knockout drops."

Mint Julep A tall drink made by breaking lump sugar in a glass, adding sprigs of fresh mint, bruised to release aromatic flavor, then filling the glass with crushed ice and whiskey.

Noggin A ¼-pint measure associated with beer.

Nuits-St.-Georges A full-bodied red burgundy named for the Côte de Nuits commune where it is produced.

Old Fashioned A cocktail made from 1 jigger of rye, ½ jigger of water, a teaspoon of sugar, and a few dashes of aromatic bitters. These are stirred over ice and served garnished with an orange slice and maraschino cherry.

Orange Blossom A cocktail made from 1 jigger of gin, the juice of ½ orange, and ¼ teaspoon sugar. These are shaken with ice and strained into the glass.

Ouzo An anise-flavored Greek liqueur that turns milky-white when diluted with water.

Pernod An aperitif named for its inventor, Henri-Louis Pernod, who opened his factory in the Jura town of Pontarlier in 1797. Flavored with aniseed, Pernod was produced as a non-toxic substitute for Absinthe.

Perry The fermented juice of pears, sweetened and filtered. There are both still and sparkling varieties.

Pinot Noir An important black grape used in red burgundy and champagne.

Port Wine named for its shipping port, Oporto, and made from grapes grown in the Upper Duoro region of Portugal. Port is fortified at vintage time. Before fermentation has converted all the grape sugar to alcohol, brandy is added for further fermentation. "Vintage" port designates wine shipped two years after production and bottled in England; "Tawny" port indicates a blend that is matured for a number of years before bottling; "Ruby" port is a cross-blend of the two types.

Porter A weak stout.

Posset Sweetened hot milk curdled with ale and flavored with nutmeg. The drink was popular in England from the Renaissance through the eighteenth century.

Punch Historically linked to the British acquisition of Jamaica in 1655, the drink was originally a mixture of rum and water, sweetened with sugar and flavored with lemons and oranges. Eighteenth-century versions add brandy and tea, elaborating on the drink, which was customarily mixed with ceremony at the table. Modern punches include various spirits, such as gin and vodka, liqueurs like Cointreau, kirsch, and Triple Sec, wines, and flavored sodas.

Retsina A white greek wine flavored with tree resin from the fermentation casks.

Rhenish The English name for Rhine wines from the earliest days of their import to the eighteenth century, when the term "hock" replaced it.

Rob Roy A cocktail made from ½ jigger of scotch and ½ jigger of sweet vermouth, flavored with a few dashes of aromatic bitters.

Rosé Pink wines, both still and sparkling, which get their color from the natural dye in fermenting black grape skins. Less desirable versions are colored with cochineal.

Rum A spirit distilled from molasses or fermented sugarcane. Jamaican rum, a dark and rich variety, is the standard. There is also a dry light Cuban version and several different flavored, aromatic rums from Puerto Rico, Trinidad, and Barbados.

Rye A grain spirit distilled from at least 51 percent rye.

Sack A corruption of the word "sec" meaning dry. This dry, amber wine is sometimes sweetened with sugar. The kind called "Sherris-sack" comes from Cádiz or Jerez; that called "Canary-sack" from Tenerife and Málaga. The wine, a favorite in Elizabethan England, has gone through a revival as an aperitif.

Saint-Émilion Wine from Bordeaux-region vineyards near the river Dordogne above Libourne and near the medieval town for which it is named.

Saint-Georges A red burgundy from Saint-Georges commune vineyards on the Côte d'Or.

Saint-Julien A claret from the Gironde Département that is the first growth of a Saint-Émilion.

Sake Japanese rice beer produced by two consecutive fermentation processes. Because it is non-carbonated it is often mistakenly thought of as a wine. The name is sometimes believed to refer in origin to Osaka, a city famous for the drink, which is served warm in small bowls.

Sauternes A rich, sweet white wine named for its commune, which adjoins the Graves district.

Sauvignon An important white wine grape.

Sazerac A cocktail made from bourbon, absinthe, bitters, and sugar, stirred with ice, strained, and served with a twist of lemon.

Schnapps A general term for various distilled liquors.

Scotch Since the early nineteenth century, the most popular grain spirit. Originally brewed by lairds for their clans and prohibited for commercial production until 1814, the whiskey is made from barley that is malted, mashed, fermented, and then distilled. It must be matured in oak casks for at least three years before sold. Different flavors are acquired by blending of malted and plain grain spirit as well as variations in grain, water, and casks.

Sherry Wine made from grapes of Spain's Jerez district, to which brandy has been added for a second fermentation process. The final product blends wines of differing maturity. The range is from pale dry (Amontillado, Fino, Vino de Porto, and Manzanilla) to dark and sweet (Oloroso, Amoroso, and Brown).

Shrub A drink popular in the eighteenth century, made from citrus juice, sugar, and rum.

Sidecar A cocktail made from ¾ ounce Cointreau or Triple Sec, 1 ounce brandy, and the juice of ½ lemon, shaken with ice and strained.

Sloe gin A cordial made from gin flavored with blackthorn fruit "sloes." This is most frequently used for a Sloe Gin Fizz made by adding the juice of a lemon and 1 teaspoon sugar to 4 ounces of sloe gin in a water glass filled with ice. Club soda or seltzer is added to fill the glass.

Stinger A cocktail made from 1 ounce white crème de menthe liqueur and 1 ounce brandy. These are shaken with ice and strained.

Stirrup cup Last or parting glass.

Stout The darkest ale or beer, which is sweet from the kilning time and long roasting of the malt.

Syllabub Hot but not curdled drink made from whole milk, sugar, and wine. Whipped cream is floated on top. The drink was popular from the Renaissance through the eighteenth century.

Tea Attributed to Shen Nung, 2737 B.C., the habit of drinking this infusion of plant leaves was brought from China to Japan in A.D. 800. There it was used as a medicine for five hundred years until green tea was developed as a beverage. While *Camellia* grows wild in the orient, three main varieties are cultivated. The multistemmed *China* bush reaches nine feet, endures cold, and can be grown in such high altitudes as Darjeeling and Ceylon. The *Assam* single-stemmed tree grows twenty to sixty feet and has many sub-varieties. The *Cambodian* tree grows to sixteen feet and is used to crossbreed. Teas are distinguished by their production process. *Black* teas are fermented; the plucked leaves are exposed to air, rolled, heated several times, and dried over a charcoal fire. *Green* teas are unfermented; the fresh leaves are rolled, heated to inactivate the enzymes, then roasted. *Oolong* tea is semi-fermented and refers to the leaves

458 · A GLOSSARY OF DRINKS

of a special plant of South China and Taiwan. Teas are graded by leaf size: congou, orange pekoe, pekoe, and souchong. The word *tea* is a corruption of the Ammoy-dialect Chinese ideogram pronounced "tay." Brought to Europe in the seventeenth century, tea was widely adopted and often pronounced in the original manner of the Chinese.

Tent English corruption of the Spanish "tinto," meaning dark-colored. It refers to a deep-red Spanish wine of particularly low alcoholic content that was used for the sacrament.

Tequila A spirit distilled from mescal, the sap (pulque) of *Agave* plants.

Toddy In the tropics, a cold drink made from fermented palm sap; in northern countries, a hot drink made with sugar, lemon, hot water, and spirits such as rum or whiskey.

Tokay Wines from the Hungarian village of Tokaj, Hungary, near the Carpathian Mountains.

Tom Collins A tall drink made originally with 1 jigger of gin, 1 teaspoon powdered sugar, soda or seltzer, and lime or lemon juice to taste. It is now commonly made with gin and a prepared soda mix named after the drink. The **John Collins** is a similar drink made with Holland gin.

Tonic water A carbonated soda water flavored with quinine, an alkaloid of the cinchona bark. Originally introduced to bring down tropical fevers, it is now a widely acquired taste, usually mixed with gin or vodka.

Valpolicella An Italian red wine produced in Verona province from Corvina, Rondinella, and Molinara grapes.

Vermouth White wine flavored with alpine herbs that are cooked in the wine during the process. The dry variety is also called French vermouth. A darker and sweeter variety is also called Italian vermouth.

Vodka A colorless, unaged spirit distilled from potatoes or malted grains such as rye and wheat.

VSOP "Very Special Old Pale," a designation of brandy aged twenty to twenty-five years.

Wassail A spiced beer or wine traditionally associated with Christmas, New Year's, and Twelfth Night. From the Anglo-Saxon *waes hael,* meaning "to your health."

Whiskey A generic term for grain spirit distilled from barley, rye, corn, or other cereals.

Whiskey Sour A cocktail made from 1 jigger of whiskey shaken with the juice of ½ lemon and sugar to taste. The drink is diluted with a splash of club soda or water and garnished with an orange slice and maraschino cherry.

Zinfandel The most widely grown varietal grape of California. Its origin has eluded researchers.

PERMISSIONS